A LITTLE BOOK OF SELF CARE

# SLEEP

A LITTLE BOOK OF SELF CARE

# SLEEP

HARNESS THE POWER OF SLEEP FOR
OPTIMAL HEALTH AND WELLBEING

PETRA HAWKER

**Senior Editor** Rona Skene
**Editor** Ian Fitzgerald
**Senior Designers**
Karen Constanti, Collette Sadler
**Designer** Nicola Erdpresser
**Editorial Assistant** Kiron Gill
**Senior Production Editor** Tony Phipps
**Production Controller** Denitsa Kenanska
**Jacket Designer** Amy Cox
**Jacket Co-ordinator** Lucy Philpott
**Managing Editor** Dawn Henderson
**Managing Art Editor** Marianne Markham
**Art Director** Maxine Pedliham
**Publishing Director** Mary-Clare Jerram
**Special Sales Creative Project Manager**
Alison Donovan

**Illustrated by** Nomoco

First published in Great Britain in 2019 by
Dorling Kindersley Limited
DK, One Embassy Gardens,
8 Viaduct Gardens, London, SW11 7BW

**DISCLAIMER** see page 144

A CIP catalogue record for this book is available
from the British Library.
ISBN: 978-0-2414-1037-0

Printed and bound in China

**For the curious**
**www.dk.com**

# CONTENTS

# FOREWORD

---

I became a sleep specialist almost by accident. I qualified as a counsellor and hypnotherapist 27 years ago, opening my first practice while continuing to study for my PhD. Over the years, clients began to tell me that not only were they feeling better about the problem that had brought them to me, but their sleep had also improved dramatically. This fascinated me: I began to delve more into the world of sleep disorders.

Then a few years ago, I joined the team at The London Sleep Centre as a sleep psychotherapist working with severely sleep-deprived patients, resulting from trauma, stress, depression, anxiety, and lifestyle issues. It has been a challenging but hugely rewarding experience, and I wanted to bring some of what I have learnt to you in this book.

In my clinical work, I see people who suffer the most debilitating sleep conditions: many are convinced they will never sleep properly again. I work with these patients to re-program their habits, building new behaviour patterns – and crucially, a positive mindset that their problem can and will be resolved.

Sleep research is constantly evolving, and we now know how vital sleep is in supporting both physical and mental health. It has been estimated that at any time, up to half of UK adults are experiencing some sort of sleep issue. Unfortunately, today's culture of 24-hour digital connectivity isn't very conducive to sleep!

But there are plenty of ways in which we can help ourselves to better rest. Some lifestyle changes are small and easy to make; others can take time and patience to bed in. My hope is that this book will help you understand better what sleep is, and to offer positive, proven strategies to help if you are having problems.

The first chapter sets out the basics of sleep – what it is, why we need it, and how it benefits us. And if you've ever searched online for help, you'll have come across that mysterious phrase "sleep hygiene" – here you'll find an explanation and an easy-to-follow checklist to use as a foundation for your healthy sleep routine.

In the second chapter, you'll find more than 40 strategies to combat a range of problems. To make it easy to find what you need, they are split over three sections covering the main areas in which we experience sleep problems; our environment, our minds, and our bodies. The emphasis is on empowering you to take control of the situation: to help yourself in every way you can, and to know when to seek help for problems that you can't fix alone.

Writing this book has been a pleasure and a privilege. I hope it will be a helpful resource whether you are struggling to sleep well, or simply wish to ensure you get the most from your daily rest.

**Petra Hawker**

# WHY SLEEP MATTERS

# THE BENEFITS
# OF SLEEP

Sleep is beneficial for mind, body, and spirit. Getting the right amount of good-quality sleep is one of the most effective self-help measures you can take to maintain or regain your health and boost your sense of wellbeing. Here are just a few of the advantages that sleeping well can bring you.

**PHYSICAL BOOST**
While you sleep, the body builds and repairs itself by renewing cells. Sleep replenishes energy and boosts your immune system, making you less susceptible to infection and viruses.

**WEIGHT CONTROL**
Research shows that poor sleep can disrupt hormones that regulate feelings of hunger and fullness. Sleeping well can help you manage food intake and control your weight.

**A CLEAR MIND**
Your brain performs vital housekeeping as you sleep, recording and sorting through the daily data we encounter, then storing it as short- or long-term memory.

**EMOTIONAL BALANCE**
Sleeping well lifts your mood and strengthens emotional resilience. There is also evidence that it can reduce the risk of depression and other mental illnesses.

**HEALTHY HEART**
Sleeping less than six hours a night increases the risk of heart disease. Deep sleep lowers both your heart rate and blood pressure, which may be factors in reducing this risk.

# THE STAGES
# OF SLEEP

For something so essential to our health and wellbeing, sleep remains something of a mystery to scientists. There is much still to understand about precisely why we need it, but we do know more now about the process itself.

## WHY DO WE SLEEP?

The short answer is that we don't know the precise function of sleep – it might be to allow body and brain to perform vital maintenance and repair, flush out toxins that have accumulated, or make and consolidate memories.

## HOW DO WE FALL ASLEEP?

As natural light fades, our brains begin to produce melatonin, which makes us feel sleepy (see pages 14–15). The longer we are awake, the more we feel sleep pressure, which is the biological urge to sleep, caused by various physical and hormonal factors.

## FOUR STAGES OF SLEEP

When we sleep, we pass repeatedly through different levels. During non-rapid eye movement (NREM) stages, we gradually sleep more deeply, but can still move our bodies around. In rapid eye movement (REM) sleep, our muscles are temporarily paralysed as we dream. It takes around 90 minutes to complete a cycle of all four stages. As the night goes on, the REM stage gets longer.

## WHY DO WE WAKE?

As it gets lighter, the body's production of melatonin tails off. Also, after a long sleep, sleep pressure is at its lowest. With no biological urge to sleep, we awaken naturally, feeling rested.

## STAGE TWO (NREM)
By this stage, you are properly asleep. Your brain activity decreases further, breathing becomes shallower and more even, your heart rate slows, and your body temperature drops.

## STAGE THREE (NREM)
This deep, restorative stage is also called slow-wave sleep. Brain activity and blood pressure are low. If you are woken up from this state, you feel confused and disorientated.

## STAGE ONE (NREM)
This is the lightest stage of sleep, during which you can easily be awoken. Your muscles begin to relax and brain activity slows. You may experience muscle jerks or spasms as you drift off. It's a state similar to hypnosis.

## STAGE FOUR (REM)
During rapid eye movement (REM) sleep, the eyes move rapidly from side to side under closed lids as we dream. Brain activity increases significantly, and breathing and heart rates rise.

**The sleep cycle**
We cycle through the stages throughout our sleeping time – so if we sleep for 7.5 hours, we would complete 5 cycles.

# OUR NATURAL SLEEP CYCLE

Over millions of years, evolution and the 24-hour orbit of the Earth around the Sun have combined to make the sleeping and waking patterns of all living things dependent on whether it is light or dark. Here's how – and why – it works.

The body has a built-in timekeeping system that sets all its rhythms, such as sleeping, eating, and digestion. This internal "body clock" is known as our circadian rhythm. It works in large part in reaction to the external, 24-hour clock: when the optic nerve detects daylight, part of our brain called the suprachiasmatic nucleus (SCN) produces the stimulating hormone serotonin to wake us up. When light levels fall at the end of the day, the body is prompted to produce the sleep-inducing hormone melatonin.

Through the day, we experience different levels of energy, alertness, wakefulness, and sleepiness, dictated in part by our circadian rhythms. This cycle takes 24 hours on average to complete, although some body clocks run faster or slower than the norm.

See pages 74–75 for more on working with your own body rhythms.

## THE ROLE OF HORMONES

Hormones play a key part in how and when we sleep. The two most important hormones in the cycle are:

• **Serotonin** Production of this chemical increases under exposure to natural light. It's sometimes known as the "happy hormone", as it helps stabilize our mood. It also promotes wakefulness and helps to regulate the nervous system.

• **Melatonin** When we are exposed to low-light conditions or darkness, our SCN tells the brain's pineal gland to start releasing melatonin. This hormone helps us sleep by lowering our body temperature and blood pressure, and by managing anxiety levels.

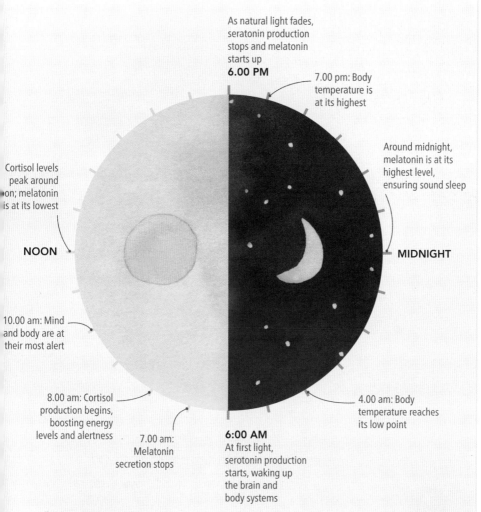

As natural light fades, seratonin production stops and melatonin starts up
**6.00 PM**

7.00 pm: Body temperature is at its highest

Around midnight, melatonin is at its highest level, ensuring sound sleep

Cortisol levels peak around ●on; melatonin is at its lowest

**NOON**

**MIDNIGHT**

10.00 am: Mind and body are at their most alert

8.00 am: Cortisol production begins, boosting energy levels and alertness

7.00 am: Melatonin secretion stops

**6:00 AM**
At first light, serotonin production starts, waking up the brain and body systems

4.00 am: Body temperature reaches its low point

## A day in the life
This chart shows how the body's production of the key sleeping and waking hormones cycle through a typical day, triggered mainly by rising and falling levels of daylight.

# YOUR BRAIN
# AND SLEEP

The brain is central to all aspects of sleep. It controls our
rhythms of sleeping and waking and the different kinds of
sleep we experience. While we slumber it remains active,
making and processing memories, and clearing old neural
pathways to enable us to learn new things.

While we still don't fully understand the
need for sleep, we now know a lot more
about the brain's role, and the functions
it performs while we are asleep.

Every human brain contains around
100 billion neurons. These are the
pathways along which information travels.
Like real roads, they are less busy at night
– and often populated by work crews
performing running repairs or building
new roads. In the brain's case, what these
work crews produce are called synapses.
These are the junctions that connect

neurons with one another, and there are
more of them in a single brain than there
are stars in the Universe.

**MAKING MEMORIES AS WE SLEEP**
While we sleep, our brain cycles constantly
through different phases of activity. During
the phase known as slow-wave sleep, the
mind "replays" what it has experienced
that day and evaluates whether or not it is
worth storing. This essential piece of
mental housekeeping is known as
consolidation.

" *Studies show that people who sleep immediately after studying do better at retaining what they have studied.* "

The hippocampus plays a crucial role in this process of turning short-term memories into long-term ones. Information is sent from the working-memory part of the brain to the hippocampus, where it circulates and is encoded, then sent along neurons and through synapses to the cortex, which is the long-term storage centre of the brain. The data then fires back and forth between hippocampus and cortex, until it is consolidated as a memory and moves permanently into the cortex.

## STUDY AND SLEEP

Studies show that people who sleep immediately after studying do better at retaining what they have studied. This is true of both kinds of memory used for learning; declarative memory, which is the ability to recall facts and information, and procedural memory, which enables you to remember how to do something, such as knit or drive a car. Recent research also suggests that the brain is even able to process and retain new information that it receives while we are asleep.

# WHAT'S YOUR SLEEP LIKE?

The first step to improving your sleep is to review what you are doing now. Once you have a full picture of your attitudes and habits, you can see more clearly where to make changes. Before you read the sleep strategies in the next chapter, consider these questions to help you identify things to improve or change.

## PREPARING FOR BED
**Do you have** a bedtime routine? **Do you sleep** alone, or with a partner – or pets? **Do you use your bedroom** for work or hobbies, as well as sleep? **Is your bedroom** as comfortable and inviting as you'd like?

Pages that might help you with any issues include: 30–31; 32–33; 36–37; 100–101.

## GETTING TO SLEEP
**Once you're in bed,** does it take you long to get to sleep? **Are you easily disturbed** by noise or other irritations? **Do you find** it hard to switch off your thoughts? **Do you feel** physically restless?

Pages that might help you with any issues include: 38–39; 62–63; 106–107.

## IN THE MORNING

**Do you wake up** refreshed after a night's sleep? **Do you hit** the snooze button repeatedly before you finally get up? **Do your muscles** ache when you get up?

Pages that might help you with any issues include: 46–47; 80–81; 94–95.

## AT NIGHT

**Do you often** wake up at night? **Are you a light sleeper** who wakes easily? **Are you often** awoken by indigestion or other physical symptoms? **Do you wake** up needing to go to the loo?

Pages that might help you with any issues include: 34–35; 40–41; 76–77; 114–115.

## DURING THE DAY

**Are you tired** and sluggish during the day? **Do you suffer** sudden dips in energy? **Is your daily** work or home life stressful? **Do you often** sleep during the day, because of work or family commitments?

Pages that might help you with any issues include: 60–61; 70–71; 74–75; 132–133.

---

# HOW MUCH SLEEP DO WE NEED?

We all need different amounts of sleep at different stages in our lives. Achieving the ideal sleep times is not always easy – treat the recommended times listed here as something to aim for and, with the help of this book, something that you can realistically achieve.

## NEWBORNS, INFANTS, AND TODDLERS

Babies up to 3 months need between 14–17 hours' sleep each day, with wake times between sleeps of 1–3 hours. One-year-olds need 12–15 hours, dropping to 11–14 hours for two-year-olds. More than half of a baby's sleep time is spent in active (REM) sleep.

## PRE-SCHOOLERS

Children aged 3–5 need 10–13 hours' sleep, including a daytime nap or two, decreasing in length as the child gets older. Night sleeps should be as continuous as possible. This is the age at which issues such as sleepwalking and night terrors are most common.

## PRE-TEENS

For children aged 6–12, sleeping for 9–11 hours is the ideal – with all of the sleep taking place at night. Children in this age group may begin to experience sleep difficulties as sleep-disrupting elements such as TV and digital screens, or stimulants in food and drink, come into play.

## TEENAGERS

Teens need 8–10 hours of sleep per night. Adolescents often undergo a change in their natural sleep patterns, with sleep onset delayed to 2 am or later, and wake time to between 10 am and 1 pm. This can lead to fatigue and lack of concentration during the day.

## ADULTS

Around 7–9 hours' sleep is ideal for adults aged from 18–65. Evidence increasingly points to the fact that we need to sleep for long enough every night; skimping on sleep in the week and catching up at the weekend does not make up for the ill effects of "sleep debt".

## OLDER PEOPLE

On average, older people need to sleep for 7–8 hours a night. It's a misconception that the need for sleep declines significantly with age; in fact, the amount we need diminishes only a little, but we experience less slow-wave sleep (especially men) and slightly less REM sleep.

# CHILDREN AND SLEEP ROUTINES

Most of this book is directed at adults who want to help themselves to sleep better. But it's never too early to get into good sleep habits; guide your children to make good sleep choices now and they will reap a lifetime of benefits.

## THE FIRST YEAR

For the first month or two of life, sleep occurs round the clock, as life revolves around feeding times. After this, a pattern will start to emerge. You can gradually start to adapt this pattern to the "normal" day-night cycle by exposing the baby to bright light during the day and so making wakeful periods slightly longer, then at night creating a dimmer, calmer environment. Put your baby to bed when he or she is sleepy, not when they are already asleep. It's important for babies to learn how to get themselves to sleep.

## UP TO FIVE YEARS OLD

For young children, establishing a sleep routine is as crucial as other aspects of self-care such as toilet training. Children who sleep well are happier, calmer, and more resilient.

**Avoid long naps** in the daytime, and don't have naps too close to bedtime. **Vigorous play should stop** an hour before bedtime. Reading a story together is a calming way to start winding down. **If your child is hungry**, give them a snack, but keep drinks to a minimum. Ensure they go to the toilet before bed.

> " *For young children, a sleep routine is as crucial as other aspects of self-care such as toilet training.* "

**Bedtime** should be at a set time. There always needs to be some flexibility, but consistency is key.

**Bedroom temperature** should not be too warm – around 24°C is ideal.

**Leave your child** to fall asleep on his or her own. A favourite toy or comforter to cuddle in bed is fine.

**If your child gets up at night**, lead them calmly back to bed with as little fuss and interaction as possible.

**Morning wake times** should be consistent – this is more beneficial to sleep patterns than even a set bedtime.

**PRE-TEENS AND TEENS**

Beware of "blue light" over-exposure with older children (see pages 36–37): ideally all electronic devices should be put away at least an hour before bedtime. As their body clock shifts, many teens genuinely struggle to stick to the sleep patterns dictated by school hours. It can be very challenging to impose or negotiate a sleep routine with older children and teenagers, but it's worth persevering as the predictability of good sleep can be a comfort at a time of hormonal upheavals, school stress, and peer pressures.

# SLEEP HYGIENE

The key to sleeping well is good sleep hygiene, which simply means developing helpful, healthy sleep habits – and sticking to them. Whatever the specific problem, improving your general sleep hygiene will boost your chances of tackling the issue successfully.

**KEEP REGULAR HOURS**
If you only observe one sleep-hygiene rule, make it this one. Going to sleep and waking up at the same time every day is the most effective way to keep your internal body clock on an even keel.

**BEDROOMS ARE FOR SLEEPING**
Don't engage in activities such as watching television, reading, or working in your bedroom. The aim is to train your brain to identify the bedroom as a place of sleep.

## WATCH YOUR INTAKE

Avoid alcohol and caffeine
for at least 6 hours before
bed. These interfere with
your natural sleep pattern.
Drink plenty of water during
the day, stopping a couple of
hours before bedtime.

## AVOID DIGITAL DEVICES

Stop using digital screens at
least an hour before you go
to bed. Avoid the temptation
to check emails or go online
in bed by not using your
mobile phone as an
alarm clock.

## KEEP ACTIVE

Avoid long daytime naps, as
this will disrupt your sleep
patterns – a 20-minute nap
is fine. Take moderate or
brisk exercise during the day,
and do a gentler activity,
such as yoga, in the evening.

## WIND DOWN

A winding-down routine
could include a warm bath, a
gentle walk, or a milky drink
– the key is to do the same
things in the same order, so
your mind and body develop
the habit of slowing down,
ready for sleep.

# SLEEP
# STRATEGIES

# YOUR ENVIRONMENT

All sorts of external factors influence how we rest, such as the room we sleep in, the bed we sleep on, or the partner we sleep with. Add to that our daily exposure to noise, light, and allergens, and it's clear that the right environment is vital for good sleep.

# CREATE A SLEEP-FRIENDLY BEDROOM

It's obvious but well worth emphasizing: your bedroom needs to be an inviting and comfortable place if you want to sleep well. Here's how to create a sleep space that works best for you.

## BEDROOM STRATEGIES

Try these helpful hints to make your bedroom a haven of peace:

• **Temperature** The ideal level for adults is around 17–18°C (60–65°F). If your feet get cold, bed socks can add comfort.

• **Light** Exposure to light lowers melatonin levels and can wake you up. Blackout blinds or heavy curtains are ideal, especially during summer when dawn breaks early.

• **Pets** Pets may be psychologically comforting, but they can interrupt rest if they move around a lot. Pet fur and saliva can also cause allergy flare-ups.

• **Plants** Unless you suffer allergies, plants are helpful to sleep. They boost oxygen levels and many are natural air purifiers, removing carbon dioxide and other toxins from the environment.

## BEDS AND BEDLINEN

Get these three elements right to enjoy a satisfying night of slumber:

• **Mattress** When shopping for a mattress, try out as many as you can lying in different positions, to assess the right level of firmness for you. A mattress topper can add comfort to an existing mattress that's too hard.

• **Pillows** Make sure pillows are the right height, so your head is level with the rest of your body. If your head is too high, the curving of the spine can restrict blood flow to the brain.

• **Blankets and duvet** A weighted blanket or heavier duvet helps to create a cosy, safe place that can help alleviate insomnia and anxiety.

# SHARING YOUR SLEEPING SPACE

Sharing a bed is not always easy. If you and your partner have different sleep needs it can result in one or both of you not getting enough good-quality sleep. Here are some tips to help avoid restless nights.

It's inevitable that you and your partner won't agree on every aspect of your sleep routine. To avoid turning your bedroom into a battleground, try to compromise wherever you can – and when that isn't possible, it might be easier than you think to change an ingrained habit or preference. There will be some trial and error involved, as you experiment with what works and what doesn't, but stick with it. In the long run you'll both be happier – and sleep better.

## DIFFERENT BEDTIMES

If you and your partner are different genders, remember that women and men generally have slightly different circadian rhythms (see pages 14–15), with women going to sleep and waking up earlier than men. If you go to bed at different times, it can be hard to avoid disturbing the sleeping partner. Eye masks and earplugs (which can also help if your partner snores) are helpful, although they can take some getting used to.

## ARRANGING YOUR SPACE

When planning your shared sleeping space, some things to consider are:

**Bed** Your bed should be the biggest you can afford, and fit in your space – preferably a super-king size. A standard

*"Women and men generally have different circadian rhythms, with women going to sleep and waking up earlier than men."*

double bed is simply not big enough for two average-sized adults to enjoy good-quality sleep.

**Mattress** If you are buying a new mattress, spend as much time as you can testing them together. Dual-firmness mattresses are available, but are expensive. You can add a mattress topper to your existing mattress, to achieve a firmness you can both live with.

**Duvet** Buy two single duvets of different weights if you can't agree. Put them inside one king-size cover, so they won't come apart.

**Temperature** If you have central heating in the bedroom, keep it at a constant moderate temperature that you can both accept as a baseline. You can then add or remove bedding, or nightclothes, as required – or open a window, if that's something you both can agree on.

**Light** Invest in a night light if one of you is regularly later to bed, or likes to read in bed while the other sleeps.

**WAKING UP**

If you habitually get up before your partner, put your clothes for the following day in a different room, so that you can get ready in the morning as quietly as possible. A vibrating alarm clock on your side of the bed is less likely to disturb your partner.

# COPING WITH ALLERGIES

Trying to sleep when you suffer from allergies can be a challenge. If you think you might have an allergy, ask your doctor to test you. Once you know what you are dealing with, you can take precautions to ensure a good night's rest.

## WHAT IS AN ALLERGY?

Put simply, allergies are the immune system's reaction to any substance it deems to be harmful. The body produces a defensive antibody called immunoglobulin, which causes the physical changes that we call an allergic reaction. Symptoms can include runny nose, scratchy throat, puffy eyes, and itchy, irritated skin. These are often worse at bedtime, because the body's anti-inflammatory hormone, cortisol, is lowest at night.

## SOURCES AND SOLUTIONS

You can avoid some allergens, but minimizing their effect is often more realistic. An air purifier in the bedroom can help you sleep better.

• **Pollen** from plants causes hayfever. Antihistamines are effective, but some make you sleepy during the day.

Opt for non-drowsy antihistamines as they don't affect sleep patterns. If you can, stay indoors when pollen is high. If you do go out, shower when you get home to rinse pollen from skin and hair, and change your clothes.

• **Animals'** saliva, fur, and dead skin cells (dander) can all trigger allergic symptoms. It's best to keep pets right out of the bedroom.

• **Dust** mites irritate the skin, causing dermatitis. They live in soft furnishings and bedding, so vacuum your home and bed thoroughly and regularly. Consider buying mite-proof covers for mattresses and pillows.

• **Mould spores**, which cause blocked noses and puffy eyes, can be tackled by airing your home – open the windows, install an extractor fan in the bathroom, and use a dehumidifier to dry out damp rooms.

# MANAGING YOUR DIGITAL DEVICES

Melatonin is known as "the sleep hormone" – it slows the body down, ready for a night's rest. However, the light emitted by devices such as tablets and phones can reduce the body's melatonin production by up to 25 per cent. Here's how to avoid the so-called blue-light effect.

### WHAT IS BLUE LIGHT?

The intense, short-wave light that glows from the screens of hardware such as TVs, computers, tablets, and mobile phones is known as blue light. Melatonin production is triggered by low light – the darker it gets, the more you produce, readying you for sleep. All light suppresses melatonin, but blue light is especially stimulating – receptors in your retinas absorb blue light more than other colours.

### OFFSET THE DAMAGE

There are strategies that can help minimize the effects of blue light:

• Expose yourself to sunlight during the day. This will boost your serotonin levels, and improve your mood.

• When using digital devices in the evening, turn down the brightness.

• If it is an option, turn devices to the "night" setting, which sets displays to warmer red or green tones.

• Bright white LED bulbs are energy-efficient, but emit blue light, so avoid them in the bedroom. Consider using low-wattage incandescent lighting, or sleep-encouraging red-light bulbs.

• Switch off all digital screens (especially mobile phones) at least an hour before bedtime, to give your body time to produce melatonin.

• Use your digital downtime for relaxing activities that prepare you for sleep, such as a warm bath.

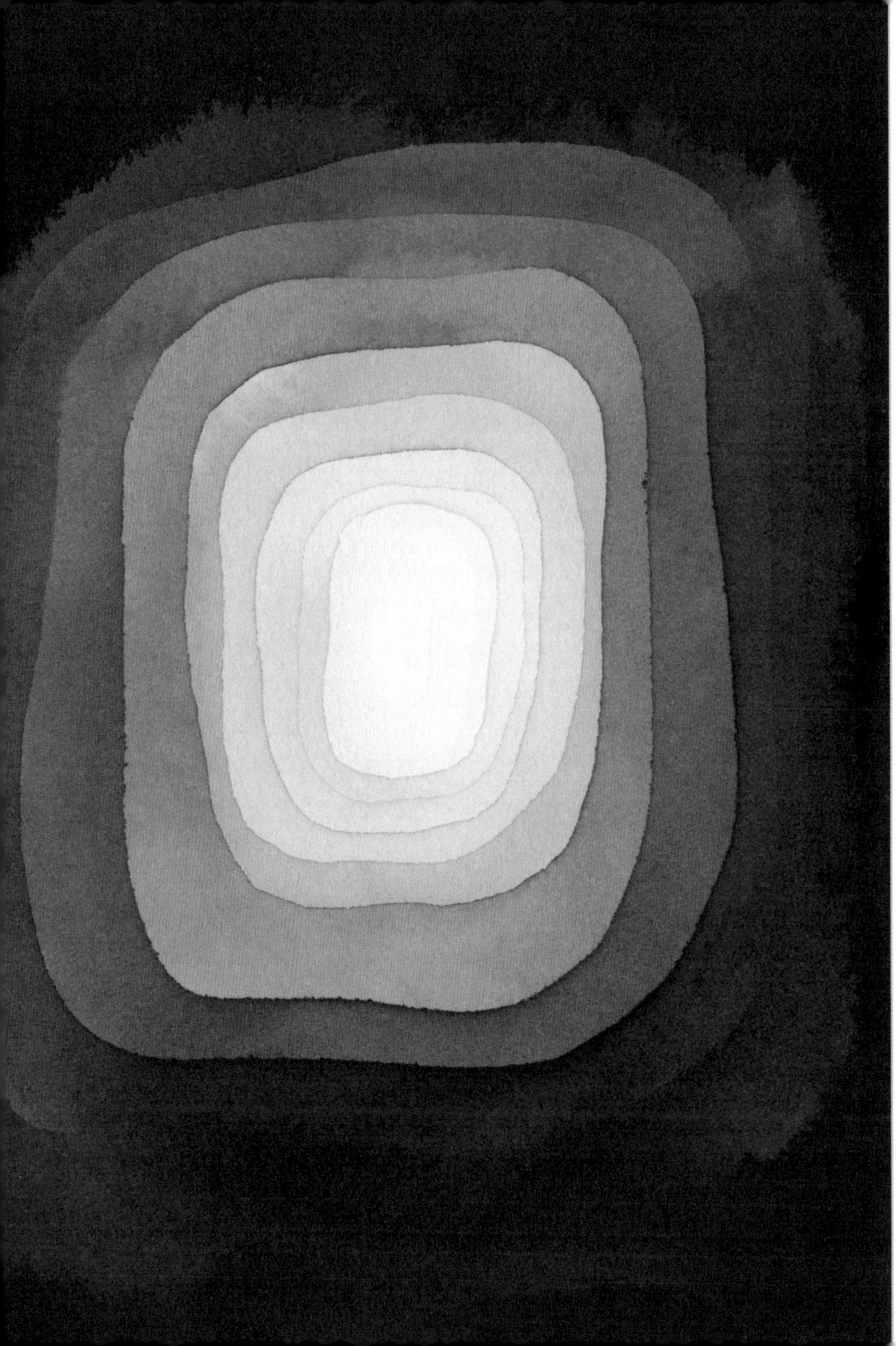

# COPING WITH NOISE

Some of us are more sensitive to noise at night than others. Unwanted sounds can be a maddening distraction, stopping you from drifting off or, if you are already asleep, they can jolt you awake. Take these steps to help you deal with the problem.

Some noise nuisances can be easily fixed, whereas others are out of your control and all you can do is minimize their effects on you.

**In the home**, a buzzing fridge can often be corrected by adjusting the thermostat. Ticking clocks can be moved, and snoring bedfellows encouraged to address their breathing problems.

**For external noise**, close doors and windows to block out street sounds. If your budget allows, double-glazing will help. If you have noisy neighbours, consider ways to minimize the issue – carpets, curtains, and soft furnishings all absorb sound. Soundproofing walls, ceilings and floors can be an effective, if costly solution.

If you are sleeping in a new place, it can be hard to adjust to different environments and sounds. Give yourself time – your brain will adapt to the change.

## REPLACE NOISE WITH SOUND

Replacing one noise with a more acceptable one can also work. The regular, relaxing pulse of white noise or binaural beats (see pages 54–55), or even a bedroom fan, can cancel out random, annoying noises and synchronize with sleep-inducing lower sound frequencies in your brain. Listening to recorded natural sounds, such as rainfall or rustling leaves, works in the same way.

## GET YOUR MIND RIGHT

You may not have control over the noise itself, but you do have power over your reaction to it. Learning to tune out noise and eliminate the stress it can trigger is often the most successful strategy. Deep breathing (see pages 92–93), visualization (see pages 60–61), and meditation (see pages 58–59) can all be helpful techniques.

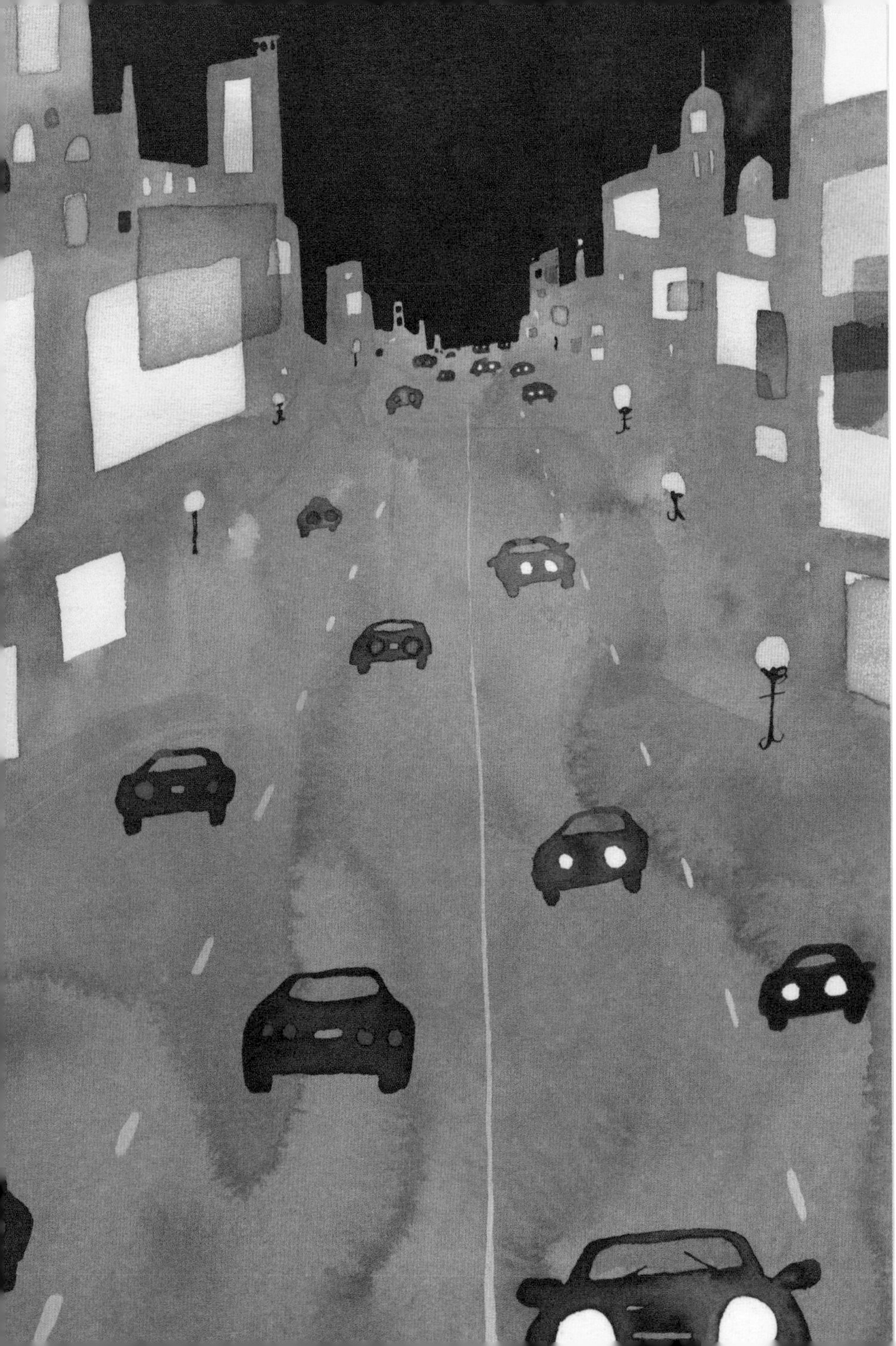

# REPAIRING BROKEN SLEEP

Waking up in the middle of the night is annoying
enough – not being able to get back to sleep
is even worse. Here's what to do to ensure you drift
off again as soon as possible.

If you are a light sleeper or live in a noisy or too-light environment, waking in the night can be an all-too-common event. If it's hard to get back to sleep once you're awake, try these ideas:

• **Keep it dark** If strong street lights are an issue, put up heavy curtains or blackout blinds. You could also try an eye mask. Some include a layer of gel and can be warmed up or cooled down, while others are scented with sleep-inducing herbs such as lavender.

• **Block out sound** If, once you wake, noise stops you drifting off again, earplugs can help. Foam earplugs are cheap, fairly effective, and designed for single use. Reusable plugs are made of silicone. They are easy to insert and remove, and comfortable to wear. Wax plugs mould to the shape of your ear, so fit better and exclude more sound than other types.

**GETTING BACK TO SLEEP**
If you do wake up, the key is not to stress about it, which only adds to the likelihood that you won't be able to sleep. Try the following instead:

• **Ignore the clock** Counting the minutes you've been awake only makes things worse.

• **Try in-bed relaxation** Starting with your toes, tense your muscles tightly, count to five, then relax. Move on to your calves, then knees, all the way up to the top of your head.

• **Get up** If the relaxation exercise doesn't work, give up for a while. Go into another room and read a book or magazine. Keep lighting and heating low, and don't turn on the TV or look at any digital devices. When you feel sleepy again, go back to bed, feeling positive about sleeping.

# YOUR MIND

The clearer and calmer your mind is, the easier it will be for you to find rest at night. In this section we'll look at ways to relieve stress, overcome anxiety, and explore how being mindful of what helps and hinders our thinking can help us sleep better.

# REWIRE YOUR THOUGHTS FOR BETTER REST

Your imagination is a powerful tool that you can use to your advantage. Visualizing a better sleep experience for yourself can prompt physical and hormonal changes that make finding sleep easier.

At bedtime, become aware of your thoughts about sleeping; are they positive or negative? Do you feel sure that you won't be able to sleep? Do you believe that you're making yourself ill by being sleep-deprived? **Be aware** that, in telling yourself you probably won't sleep, you are setting yourself up to make it true. Your stressful thoughts increase the body's levels of adrenaline and cortisol; you feel anxious and afraid, and this is not conducive to sleep.

**Change your story** The moment you start to think negatively about sleep, catch the thought and say to yourself: "Delete!" Then say the opposite to what you just thought. Simplistic as it sounds, you really can retrain thoughts in this way.

**Try visualization** (see pages 60–61). The more you do this, the more adept you will become at replacing your negative beliefs with positive, hopeful ones. Believe in the power of your mind and the possibility of change.

" *Be aware that, in telling yourself you probably won't sleep, you are setting yourself up to make it true.* "

**THINK ABOUT THERAPY**

If you have trouble turning round negativity by yourself, professional help could be the answer. Ask your doctor, or see pages 142–143 for other bodies to contact.

**Cognitive focusing** is a technique that guides you into shifting negative thoughts into positive ones.

**Systematic desensitization** is a kind of written, list-based version of cognitive focusing, where you record your adverse nighttime associations and attach positive alternatives in order to detoxify them.

**Sleep restriction therapy** trains your brain to associate bed with being a place of sleep and not restlessness. It focuses on establishing strict routines, such as set bed times and waking times.

**Cognitive behaviour therapy (CBT)** challenges your negative ideas and habits and helps you establish more positive thought and behaviour patterns (see pages 66–67).

# SWEET DREAMS, GREAT SLEEP

Research shows that having good dreams improves sleep quality and wellbeing. If you constantly wake up exhausted instead of refreshed, it could be your dreams that are responsible.

**DREAM HAPPY, SLEEP HAPPY**

If good dreams influence how we feel and help us sleep well, how do we ensure our dreams are positive? One way is to follow the advice in this book and simply try to go to bed happy! A more active technique is Imagery Rehearsal Therapy. This involves visualizing happier endings to any bad dreams you've had, usually around 15 minutes before you go to sleep. This prepares your mind for positive thoughts while sleeping.

**AVOID SLEEPING STRESSED**

When we go to bed stressed or angry, our levels of adrenaline and noradrenaline are raised. These two hormones disrupt REM sleep, which is the period in which we dream most.

This means that not only do stressed people sleep less, they dream less too, allowing insufficient time for dreams to process events, emotions, and memories, which in turn leads to even more stress. To break this cycle, try to ensure you are relaxed and calm at bedtime – for instance, by doing a yoga routine (see pages 84–85) or by meditating (see pages 58–59).

**KEEP A DREAM DIARY**

Your dreaming brain is your internal counsellor: listen to it. It's a good idea to keep a journal of your dreams. Record not just the events in your dream but how it made you feel – and how you felt before you went to sleep. Patterns will emerge that you can think about and address.

# COLOURS FOR A GOOD NIGHT'S REST

Colour affects our mood – whether it's the hues of nature, such as flowers, trees, sea, and sky or the shades we choose to wear or put in our homes. Understanding which colours lift or lower our spirits means we can use them to help us sleep.

Research shows that people are generally more stimulated by warm colours (reds and oranges) and are calmed by cooler shades (blues and greens). It makes sense to ensure that the colours you choose for your sleeping space induce a sense of peace.

A UK survey found that the three colours people reported as most restful for walls and furnishings were blue, followed by green, then yellow. Red, purple, and brown were the least likely colours to promote rest.

Colour intensity also has an effect on mood: muted, softer shades are preferable to bright, primary colours.

**GET THE LIGHT RIGHT**

There's evidence that using coloured room lighting can benefit sleep:

• **Red** light waves are known to have the least effect of all the colours on sleep-inducing melatonin. Consider using red bulbs in the bedroom, especially for nightlights.

• **Yellow** lighting boosts worry-busting serotonin production.

• **Blue** light should be avoided; it suppresses melatonin production (see pages 36–37). Children are especially vulnerable to blue light, as the light and colour receptors in their eyes are more sensitive.

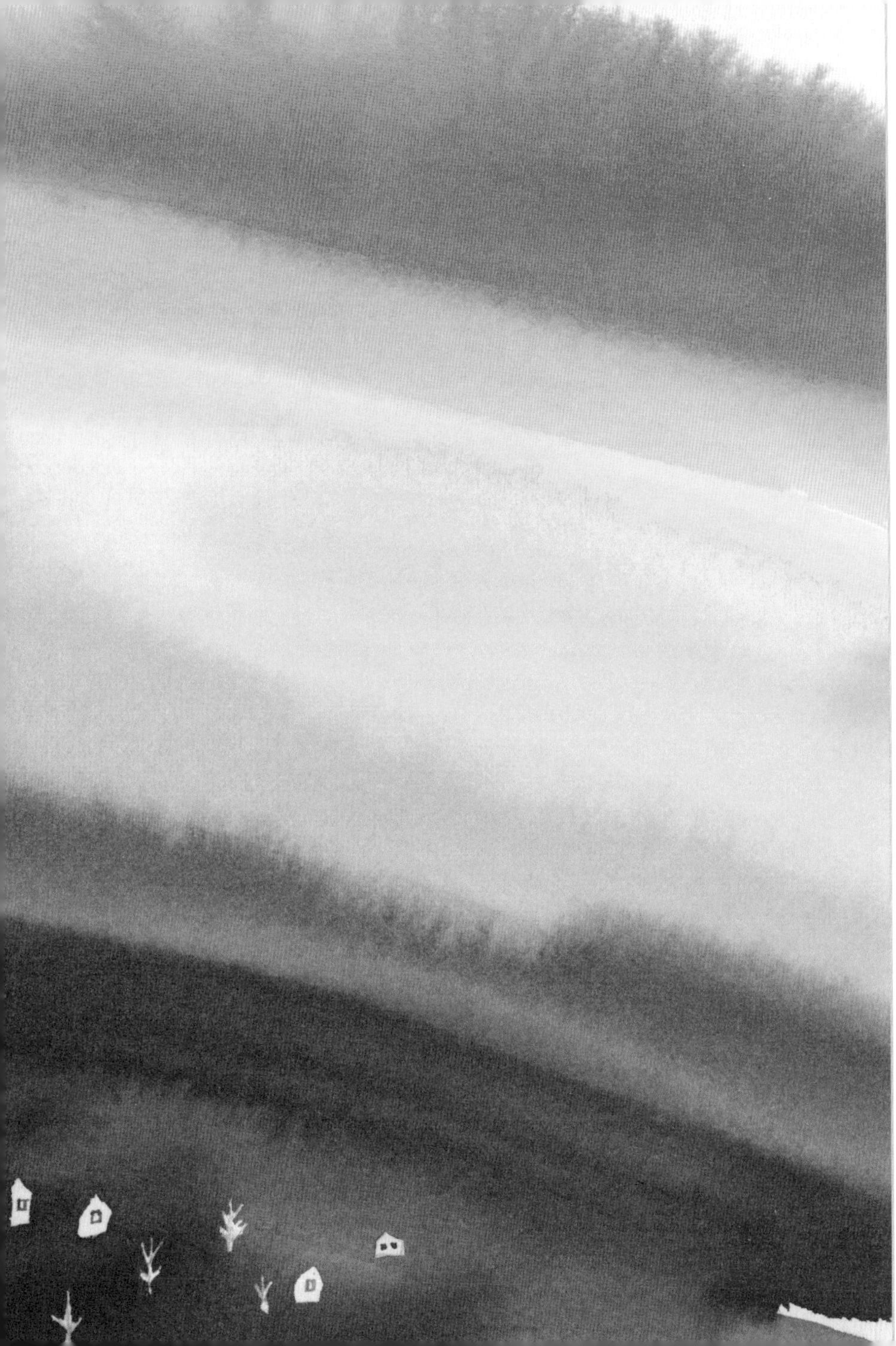

# TUNE INTO NATURE

The more we commune with nature, the more we benefit from it. Being mindful of our environment and enjoying the sensations it arouses in us helps us to live more happily, relax effectively, and sleep properly.

Enjoying nature has two great benefits for sleep. It makes us feel emotionally balanced, and it physically improves our overall health.

A good walk restores and invigorates. The key is to engage all your senses: notice scents and noises; the calming sound of waves on the shore; melodious birdsong; the heady perfume of flowers. This is mindfulness in action, using nature's rhythms and sounds to soothe your body and spirit. It's a simple lifestyle change and it is very effective. Try it: you will feel happier and find it easier to rest.

## FEEL THE BENEFITS

There are other, measurable benefits to tuning into nature. Briny, ozone-laden sea air, for example, is packed with healthy negative ions that help our bodies absorb oxygen. This in turn promotes happy-making seratonin production and raises levels of the melatonin that promotes restfulness.

Researchers have also discovered that enjoying the great outdoors helps combat a condition called oxidative stress, which, amongst other things, causes sleep problems. It's caused by a lack of healing antioxidants in your body's cells and an excess of harmful free radicals, and there is evidence that a more active, outdoor lifestyle can counteract its ill effects. Ways to lower oxidative stress levels include:

• **Gentle, outdoor exercise** such as nature rambles or gardening.

• **Breathing pollution-free air**, so take your calming walks in unspoilt coastal or country areas.

• **Reducing stress and anxiety**, by practising mindfulness, for example.

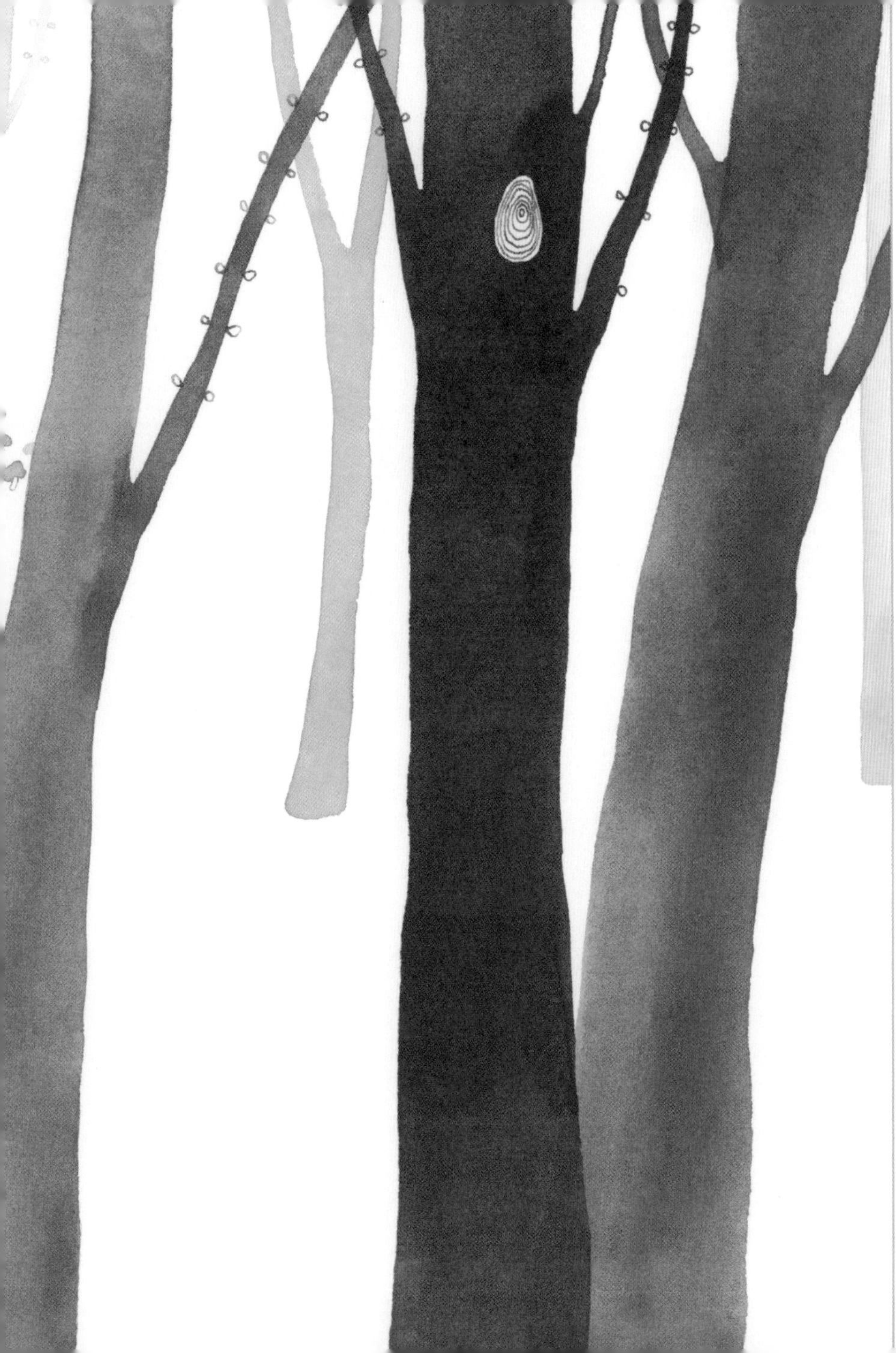

# MUSIC TO
# PREPARE FOR SLEEP

Listening to music that we like increases our body's dopamine levels. As dopamine in part regulates our emotions and helps with melatonin production, there is a positive link between music, mood, and rest.

To promote better sleep, the best time to listen to music is up to an hour before bed – but not *in* bed, which is best kept as a place of quiet.

## THE BENEFITS OF MUSIC

Numerous sleep studies have shown the benefits of pre-bed listening to music. They include:

• **Slower heart rate**, which primes the body to go into sleep mode.

• **Improved and regular breathing**, another sleep-friendly necessity.

• **Lower blood pressure**, indicating that you are in a relaxed state, and not subject to stress or anxiety.

• **Relaxed muscles** It is difficult, if not impossible, to sleep when you are physically tense.

In general, adults who listen to music before bed fall asleep faster,

sleep longer, have less interrupted sleep, and feel more rested in the morning. This is because music directly affects your parasympathetic nervous system, which helps your body ready itself for sleep.

## WHAT TO LISTEN TO

Choose something gentle and, preferably, instrumental: songs with words can be distracting. These qualities bestow further benefits:

• **Gentle, repetitive rhythms** can have a mildly hypnotic effect.

• **Low tones** at the deeper end of the sound spectrum are more resonant and calming than higher tones.

• **A slow tempo**, around 60–80 bpm (beats per minute). Slower is better, as your heart rate will synchronize to the beat of the music.

# USING SOUNDS
# TO FALL ASLEEP

It may seem a strange idea that exposing yourself to noise will help you sleep, but there's plenty of evidence that, for some people, listening to certain sounds can make sleep easier to achieve and maintain.

If you find it difficult to drift off once you get into bed, you may be tempted to listen to some music. Don't. Music before bed can be helpful in getting you in the right frame of mind for sleep (see pages 52–53), but when you're settling down to sleep, it's too intrusive and stimulating. Music with lyrics can be especially, especially when the words lodge in your psyche as a maddening "earworm". In a nutshell, music is too emotionally engaging to nod off to. What you need is

something tuneless and impersonal. What you need is noise.

## SOOTHING SOUNDS

The brain is primed to react to sound; an unexpected, loud, or high-pitched noise triggers the brain's salience network. This puts the body on a high state of alert, where sleep is the last thing on the agenda. Low-pitched, steady sounds, on the other hand, encourage the brain to disengage and enter the "wandering" state that's

" *Low-pitched, steady sounds encourage the brain to enter the 'wandering' state that's conducive to sleep.* "

conducive to sleep. Some of the best sounds for sleep include:

**White noise** This is a combination of all the sound frequencies our ears can detect. The tones all merge into a single, soothing sound. You can buy white noise machines, download an app, or even create your own with a household appliance such as a fan or air purifier.

**Natural sounds** The soft pitter-patter of rain or the rhythm of lapping waves have known calming effects. Their natural, repeating rhythms signal to our brains that we are in a safe environment and that we can go to sleep. If you aren't lucky enough to live by the ocean, you can buy recordings of a variety of natural sounds.

**Binaural beats** This system uses two pulsing soundwaves of different frequencies, one in each ear. This slows your brainwaves, creating a sleep-inducing, meditation-like effect. Downloadable sound files are widely available on the internet.

# CREATING A
# LIFE BALANCE

The way to a good night's sleep is achieving a good life balance between work commitments, family time, leisure, and rest. It can be a challenge, as so many of us lead time-pressured lives – but here are some tips to finding the equilibrium you need.

## KEEP WORK SEPARATE

Long working days can lead to late nights and ever-earlier starts – a sure recipe for disrupted sleep. It's important to establish boundaries between your job and personal life – especially now that email and social media make us contactable at all hours. It can be helpful to instigate a small end-of-day ritual to mark the end of work and the beginning of downtime.

## FIND TIME TO THINK AND FEEL

Use the time you've won back to enjoy a range of different pursuits and hobbies. Engaging both the "thinking" and "creative" parts of the brain helps to maintain a sense of balance and calm.

## FOCUS ON THE POSITIVE

You can boost your mood by using a simple visualization technique. Remembering a happy or fun event, and calling it to mind in detail will trigger the production of feel-good hormones serotonin and dopamine. Do this before bed, or any time you feel you need to lift your mood.

## MAKE CONNECTIONS

Feeling connected to loved ones and sharing feelings with them will make you feel more supported and lead to fewer worries and doubts at bedtime. Engage in more conversation – at meal times, try agreeing to turn off your TV and mobile devices so you can catch up with each other properly.

# MINDFUL MEDITATION

Meditation – focusing the mind in order to achieve calm – benefits both your physical and mental wellbeing, so not surprisingly, it's a proven aid to better sleep. Getting the hang of meditation can take a bit of practice, but the more you do it, the easier you'll find it. Don't worry about how well or badly you are doing – simply accept the experience for what it is.

## 02

With your eyes closed, breathe deeply for a few moments. Focus on your breath and become aware of your body. Relax your muscles and your mind.

## 01

Turn off any phones, TVs and radios. Find a quiet place and sit comfortably, either on the floor or in a chair with your bare feet on the ground.

### NEED TO KNOW

**BENEFITS** Calms the mind; slows heartbeat and regulates the nervous system; reduces stress and anxiety.

**TIME** Daily; 1–2 minutes at first, rising to 15 minutes after 2–3 weeks. Practise daily for maximum benefit, especially before bed.

## 03

If thoughts keep occurring to you, simply acknowledge them, then let them drift away like a passing cloud. Don't force anything; be gentle with yourself.

## 04

Observe any sounds or smells around you, and whether you feel hot, warm, or cold. Be aware of the connection between your body and the ground. Keep focusing on your gentle breath.

## 05

When you are ready, gently open your eyes and remind yourself how it felt to be focused and calm. Take those feelings with you as you continue your day.

## 01

Lie down and close your eyes.
Take three deep, slow breaths
in and out. Visualize
energizing, white light entering
your body as you inhale, and
negative energy leaving your
body as you breathe out.

## 02

Continuing to breathe consciously,
tense your toes for at least 20
seconds, then relax them. Repeat
with your calves, then your thighs,
buttocks, back, hands, arms,
shoulders, chest, neck, and finally
your face. Imagine a wave flowing
up your body, relaxing
you more with each breath.

# DAILY RELAXATION ROUTINE

Relaxation is the antidote to anxiety – slowing the heart,
lowering blood pressure, and releasing feel-good
hormones. This exercise, inspired by the spiritual
practice of Reiki, induces a calm that will last through
the day to bedtime.

## 03

Now visualize yourself descending steps into a beautiful summer garden, filled with flowers. Be aware of the blue sky and the gentle sun on your face. Sit on a bench under a tree, close your eyes, and clear your mind.

## 04

Allow yourself to feel gratitude for all the good things in your life; believe in your own capabilities, and feel yourself grow stronger. Stay there as long as you feel you need to.

## 05

When you're ready, imagine walking back through the garden and up the steps, feeling calm and relaxed. Open your eyes and continue with your day.

### NEED TO KNOW

**BENEFITS** Quiets the mind and releases tension in the body; creates a day-long calm that makes sleep easier.

**TIME** For as long as you feel necessary; ideally not less than 10 minutes. Practise daily, ideally in the morning.

## 01

Lie comfortably in bed. Begin by slowing and deepening your breathing. Don't force anything – just focus on the rise and fall of your chest and how your body feels. Now breathe in to a count of 7, pause for a count of 5, then exhale to a count of 7. Continue for a few minutes.

## 02

Return to breathing slowly and deeply. Now tense and relax your muscles, starting with your toes, and working up to your head. Tense the muscles on your in-breath, and relax them as you exhale.

# RELAX INTO SLEEP

No matter how tired we feel, many of us have trouble falling asleep because we can't switch off the thoughts that continue to swirl round in our minds. Try this simple routine in bed to quieten your mind and help you drift off.

**04**

Now start counting backwards from 1,000 with each complete breath. This routine activity allows the brain to disengage – you should be asleep long before you complete the countdown!

**03**

Visualize a favourite place or happy memory. Really focus on the sensory details; for example, the sound of lapping waves, the briny smell of seaweed, or a gentle breeze on your face. The more vividly you imagine the details, the more relaxed you'll feel.

**NEED TO KNOW**

**BENEFITS** Helps to slow thought processes and relax the body, encouraging sleep.

**TIME** Spend as long as you want on each step. Practise at night in bed, whenever you need to.

# HYPNOTHERAPY FOR SLEEP ISSUES

Hypnosis is a form of relaxation. It slows brainwaves and creates an altered state of consciousness similar to when we meditate or daydream, the difference being that under hypnosis our thoughts and feelings can be guided – for instance, towards sleeping better and for longer.

If you want to try hypnotherapy, always work with a qualified professional, preferably with specific expertise in sleep issues. Sessions typically last around an hour, for an agreed number of sessions. Before you begin, the therapist will ask detailed questions about your problem, and any outcomes you wish to achieve.

Here's what to expect in a typical hypnotherapy session:

**Close your eyes** – the therapist will talk you through a physical and mental relaxation process to prepare your mind and body for the therapy.

**As you relax**, your therapist will ensure you are comfortable, calm, and willing to continue.

**Using visualizations** and a gentle tone of voice, the hypnotherapist will put you in a deeper state of relaxation.

**The therapy begins** – the therapist starts to recommend ways for you to take back control of your sleep. They may set out a routine for you to follow, or make specific suggestions. In your altered state, you absorb these on a deep level, so that when you return to normal consciousness, they become actions and processes you

"
*Research shows that the more positively you engage with the treatment, the better the outcome is likely to be.* "

undertake almost automatically and without thinking.

**Slowly, gently**, the therapist brings you back to a fully conscious state. You come back to the world positive, refreshed, and energized.

## BENEFITS OF HYPNOTHERAPY

Anyone can undergo hypnotherapy, although if you currently have any mental health issues, you should discuss this first with your therapist.

**It is a simple and safe** intervention, with no medication involved.

**It encourages you** to be proactive – research shows that the more positively you engage with the treatment, the better the outcome is likely to be.

**As a complementary therapy**, it can be used alongside conventional medical treatments with no ill effects.

**It can work swiftly** – sometimes after a single session. You should notice positive changes after a month at most.

**You can continue** the therapy after the treatment ends – many therapists will provide recordings of sessions for you to replay at home.

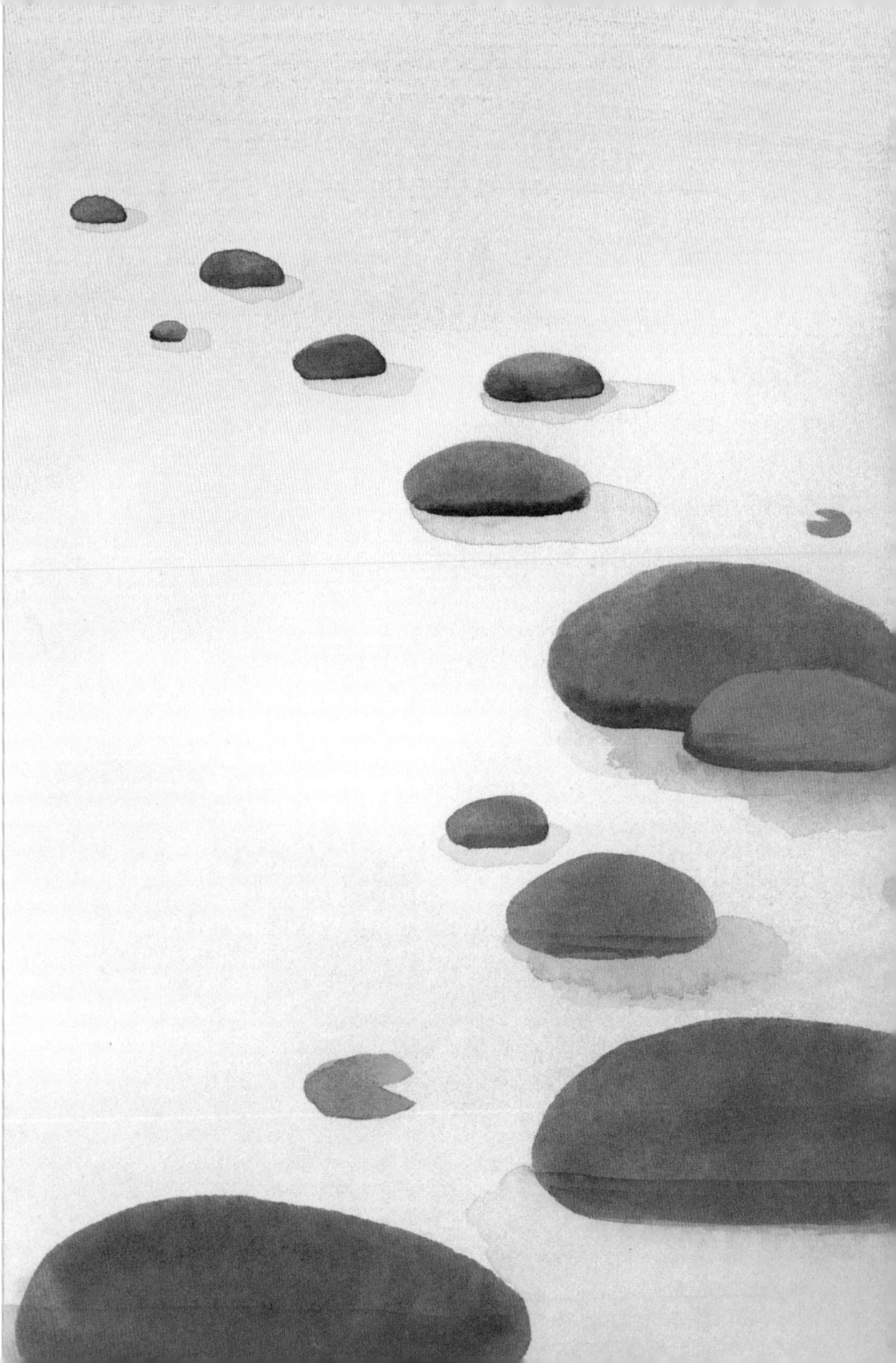

# COGNITIVE BEHAVIOURAL THERAPY FOR INSOMNIA

Cognitive behavioural therapy (CBT) is an effective way of dealing with various problems by "reprogramming" your brain. When used to treat insomnia it's called CBTi.

CBT is known as one of the talking therapies – others include counselling and psychotherapy. CBT employs a highly structured, focused, and practical approach to a specific issue, making it an effective treatment for issues such as sleeplessness. You will need a qualified therapist to guide you through what is a very detailed process. Your doctor may refer you to a practitioner, or you can find one yourself through one of the professional bodies (see pages 142–143 for resources).

## HOW DOES IT WORK?

A CBTi programme typically takes place over five or six sessions, spanning twelve weeks. CBTi works by taking your sleep problem and breaking it down into five different modules: situations, thoughts, emotions, physical sensations, and actions. Guided by your therapist, you work your way through each module, looking at how that aspect impacts on your problem, and how it interconnects with the other modules.

Piece by piece, you explore past events, physical experiences, troubling thoughts, and more, so that you build a comprehensive picture. Once you have identified the habits and circumstances that are preventing you from sleeping, you work with the practitioner to devise practical, achievable ways to overcome them. The emphasis is as much on future prevention as on resolving the immediate issue.

## 01

An hour before bedtime take out your journal, write the date, then jot down up to 5 things that are worrying you. Don't put too much detail – a brief sentence for each is fine. Leave a space after each worry.

## 02

Then under each worry write something you can do to resolve the situation. If you can't think of a solution, write a positive statement instead, such as: "I won't let this beat me," or "I can always ask my friend for advice about this."

# KEEP A WORRY JOURNAL

Sleep can be hard to find when your mind is buzzing. Try keeping a worry journal – by writing down your worries, you'll symbolically remove them from your mind. Writing longhand slows you down, giving you time to process thoughts. And by identifying specific concerns, you are well placed to come up with possible solutions.

## 03

When you have finished, read through everything you've written. Then slam the book shut with a resounding thud and put it away. You're sending your brain a message that you are done with those worries for the night.

## 04

Once you are in bed, if you find your mind drifting back to your worries, remember the satisfying sound of your journal flattening them. It's a surprisingly effective way to stop those thoughts from taking hold!

**NEED TO KNOW**

**BENEFITS** Allows space for calm reflection on your day; helps your mind to shift from "active" to "sleep" mode.

**TIME** Around 10 minutes. For maximum benefit, make the journal a nightly ritual.

**ITEMS NEEDED** A hardback journal or notebook; pen.

# MANAGE YOUR
# STRESS LEVELS

With all the complexities of life to navigate, none of us
is immune to stress. It impacts hugely on how we unwind at
night, so here's some guidance for ensuring stress doesn't
ruin your night's sleep.

Stress is our physical and mental response to pressure. It isn't always negative: stress can motivate us and promote positive change. However, if we feel powerless or overwhelmed by an event or situation, stress becomes a problem. We can't prevent stress, but managing our response to it can minimize any negative impact.

• **Massage or meditation** will unknot tight muscles and clear a cloudy mind. They both promote good sleep and lower stress-induced adrenaline and cortisol levels that can lead to heart problems and high blood pressure.

• **Exercise** is an excellent stress-buster, loosening tense muscles and releasing stress-fighting endorphins. Physical tiredness can help the mind switch off, ready for sleep.

• **Making a to-do list** for the next day can help you break down a daunting schedule into smaller, manageable chunks. Ticking off tasks through the day gives a sense of regaining control.

• **Share the burden** – if you keep stress to yourself, you can lose perspective and make the problem worse. Simply talking things over with a friend or colleague can help to lighten the load.

• **Get professional help** – if your stress makes you feel frightened or hopeless, there are therapies that can help you through. Cognitive Behavioural Therapy (CBT– see pages 66–67) can help significantly, especially following major emotional upheavals that can trigger Post-Traumatic Stress Disorder (PTSD).

# YOUR BODY

Looking after your body is key to
sleeping well. From eating healthily
and staying fit to taking natural
supplements or practising yoga, the
following pages contain helpful tips
and techniques to make sure you
stay in the right shape to sleep.

# RESET YOUR BODY CLOCK

Until the invention of artificial light, we woke up at sunrise and slept when night fell. Since then, the natural circadian rhythms have become disrupted, and many of us feel we are constantly at odds with our body clocks.

## WORK WITH YOUR TYPE

Your natural body clock is set by your circadian rhythm (see pages 14–15). Those who wake with the lark and prefer an early night are known as early chronotypes. Night owls, on the other hand, are classed as late chronotypes. Whatever our type, however, we all have to compromise to accommodate life's demands. Here are some tips for working with your type:

• Aim to wake up at the same time every day, even at weekends. This will help your mind and body realign with the 24-hour cycle.

• If possible, don't fight your type: morning larks won't thrive working night shifts, and night owls shouldn't deliver milk! Plan to do the most demanding work at the time of day you naturally feel most alert.

• It's tempting to use caffeine or alcohol as quick fixes when your sleep goes awry, but don't rely on them as part of your routine – they diminish the quality and quantity of your sleep.

• If you are a late chronotype, a daytime nap could be beneficial. A 20-minute snooze is re-energizing, without making you feel sluggish.

• Use subdued lighting in the evening – the lower the wattage, the less it will interfere with melatonin levels.

• Stick to a regular bedtime: research shows that when natural sleep rhythms are irregular, blood sugars increase, increasing the risk of illnesses such as type-2 diabetes.

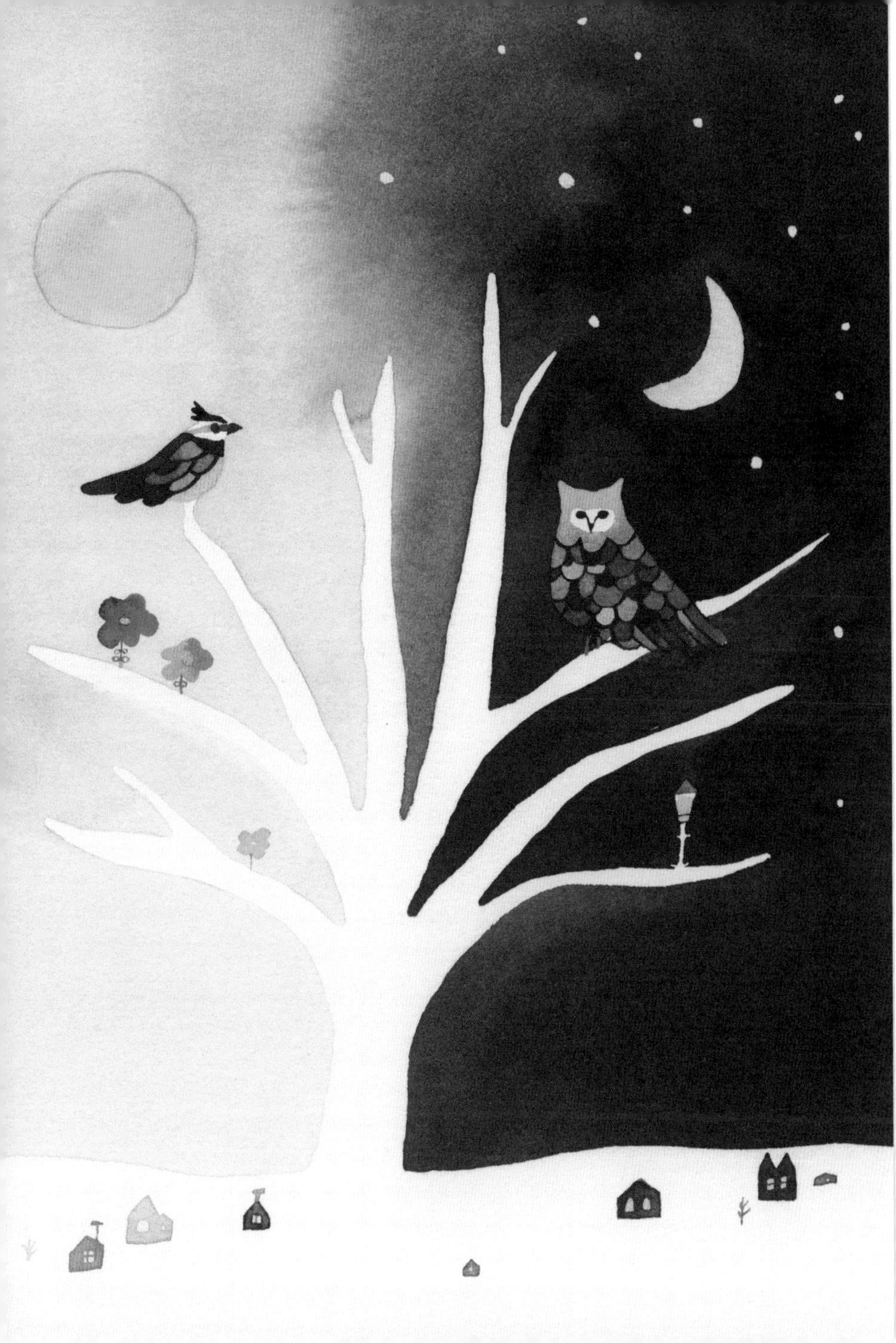

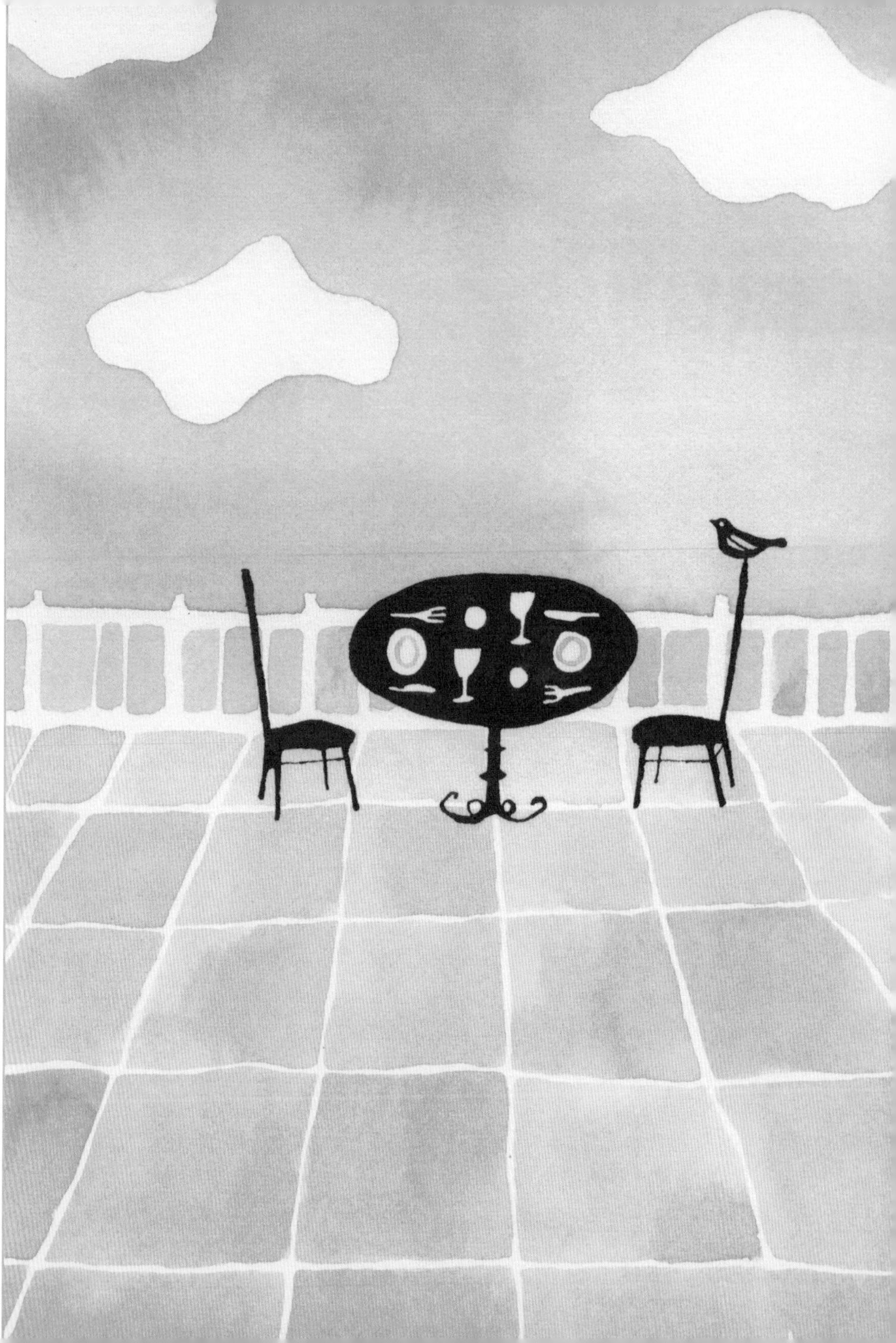

# WHAT TO EAT

Your diet has a role to play in how you sleep. Eating the right things can prepare your body for a good night's rest, especially if they contain one particular super-ingredient.

The body functions best on a balanced diet of protein, fats, and carbohydrates. In addition, there are foods that are rich in sleep-friendly substances – and others that should be avoided.

## BENEFICIAL FOODS

Foods containing the amino acid tryptophan are known to aid sleep: it helps your brain produce serotonin, one of the building blocks of good sleep.

**Tryptophan-rich foods** include: turkey and chicken; oily fish, especially salmon, tuna, and sardines; eggs; leafy vegetables including spinach; nuts and seeds, which are also loaded with healthy fats, protein, and fibre; milk and dairy foods – low-fat versions are just as helpful as full-fat.

Other foods containing ingredients that also contribute to a better night's sleep include:

• **Fruits** such as kiwi fruit and bananas contain melatonin, the sleep hormone.

• **Almonds** are rich in sleep-enhancing magnesium.

• **Lettuce** contains lactucarium, a milky substance that acts as a mild sedative.

## WHAT TO AVOID

As well as caffeine and alcohol (see pages 80–81), there are other foods best avoided at night. Try keeping a sleep/food diary to pinpoint your poor-sleep triggers.

• **Spicy foods** can cause heartburn; chilli can raise body temperature, which suppresses sleep-hormone production.

• **High-fat foods** stimulate acid production in the stomach, which can bring on indigestion.

• **Hard cheese** is rich in the amino acid tyramine, which acts as a stimulant on the brain.

# WHEN TO EAT

When it comes to sleep, when you eat is every bit as important as what you eat. Eating well, and at the right time, will set you on the road to getting your optimum amount of restful, restorative sleep.

Your body clock thrives on routine. For the best chance of sleeping well, stick to regular meal times.

**Breakfast** Avoid sweet cereals and processed white bread, which are packed with sugar. The boost that sugar delivers is followed by an energy crash that disrupts your body's rhythms. Instead, start your day with low-release energy foods combined with protein: porridge with milk, yoghurt with nuts and seeds, or wholegrain bread and cereals.

**Lunch** Steer clear of processed fast foods for lunch and opt for a mix of complex carbohydrates and protein: chicken or tofu wholewheat pasta salad, for example. Snack on slow-energy fruit or nuts, rather than biscuits or cake, as these will help you avoid energy surges and dips.

**Evening meal** Eat a balanced meal at least three hours before bed. This gives you time to digest food properly. Be mindful that the larger the meal, the longer it will take to digest. Lying down

" *Eat your main evening meal at least three hours before you go to bed.* "

on a full stomach can allow digestive juices to leak up into the throat (acid reflux), causing heartburn. Avoid high-fat foods, which stimulate acid production in the stomach, also resulting in heartburn. Digesting food also raises your body temperature, which can cause problems, because sleep hormones are suppressed when your temperature rises. .

**Late snack** If you think hunger might stop you from drifting off, a slow-energy oatcake with soft cheese, along with

tryptophan-rich warm milk is a safe bet. Avoid hard cheese – it contains a brain-stimulating protein.

**Eating in the night** Waking up hungry can be caused by low blood sugar. If it happens regularly, see your doctor – you could have one of a range of conditions that require treatment. For the occasional bout of night hunger, it's fine to treat yourself to a small snack. Avoid too much sugar or starch – try a rice cake with peanut butter, or a banana.

# STIMULANTS AND DEPRESSANTS

Certain foods, drinks, and pharmaceuticals can over-stimulate or depress our nervous systems. If we take too much, or take them at the wrong times, they can prevent us from getting the best quality of rest.

## STIMULANTS

Stimulants increase your heart rate and boost mental energy, making sleep more difficult.

• **Caffeine** can stay in your body for ten hours, so if you can, avoid coffee, tea, and energy drinks after midday.

• **Nicotine** – smokers take longer to fall asleep, sleep for an average 33 minutes less every night, and sleep more lightly than non-smokers. The best advice is to avoid nicotine. Withdrawing from nicotine can make your sleep worse, but this effect is short-lived.

• **Refined sugar** provides a rush of energy followed by a dip, so it can interfere with a sleep routine. People who consume the most sugar are also more likely to suffer interrupted sleep. Keep sugar intake to a minimum, and consume it early in the day.

## DEPRESSANTS

Although they make you drowsy, depressants can significantly affect the quality of your sleep.

• **Alcohol** reduces restorative REM sleep (see pages 12–13), so you wake up tired and sluggish. It's also a diuretic, so you find yourself having to get up to go to the toilet during the night. You'll sleep better by reducing your intake, and stopping as early in the evening as you can.

• **Antihistamines** – although they can make you fall asleep more easily, studies show they can lead to other problems such as sleep-walking. Use these drugs only to treat allergies and hayfever, not as a long-term sleep aid.

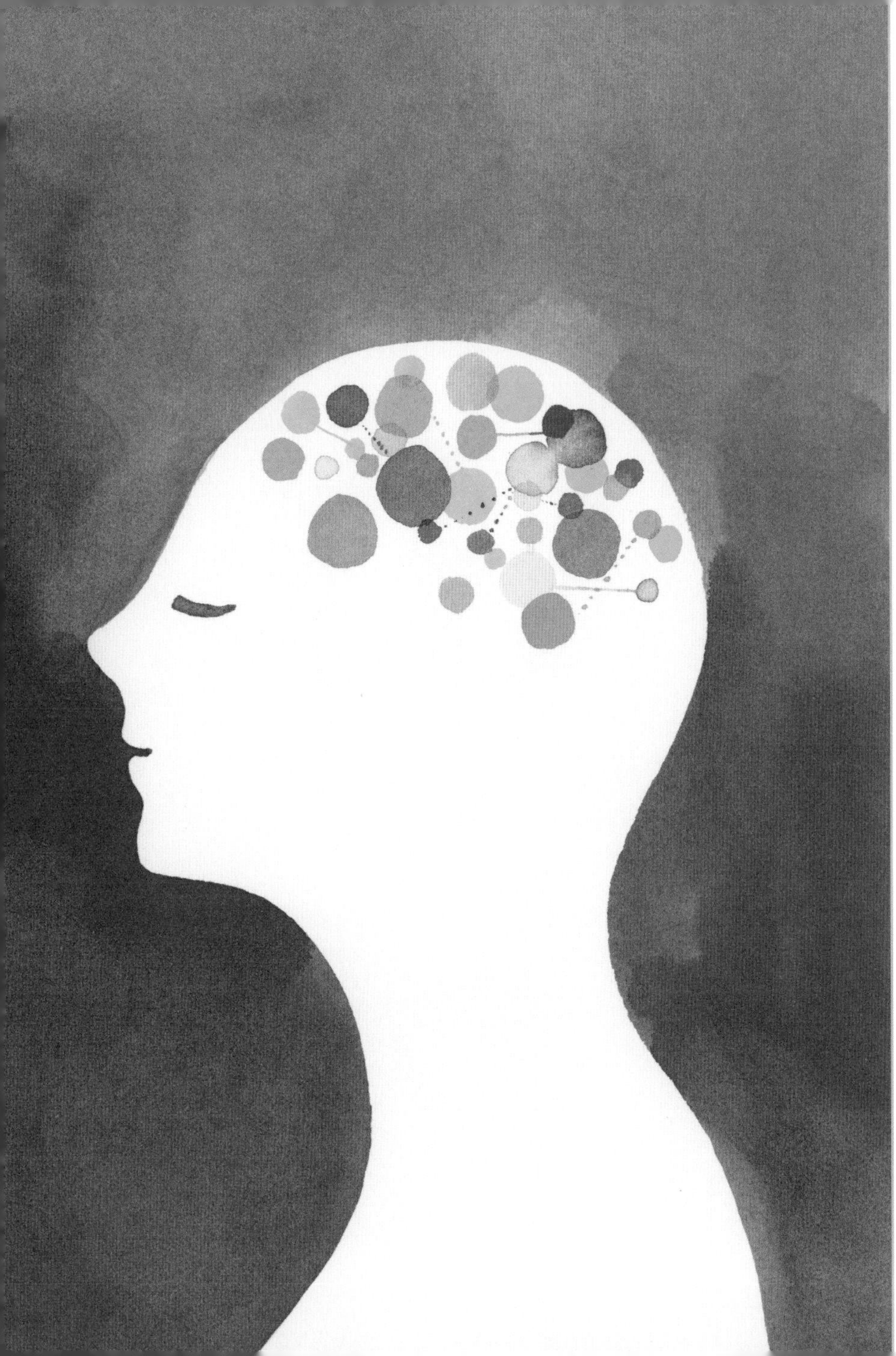

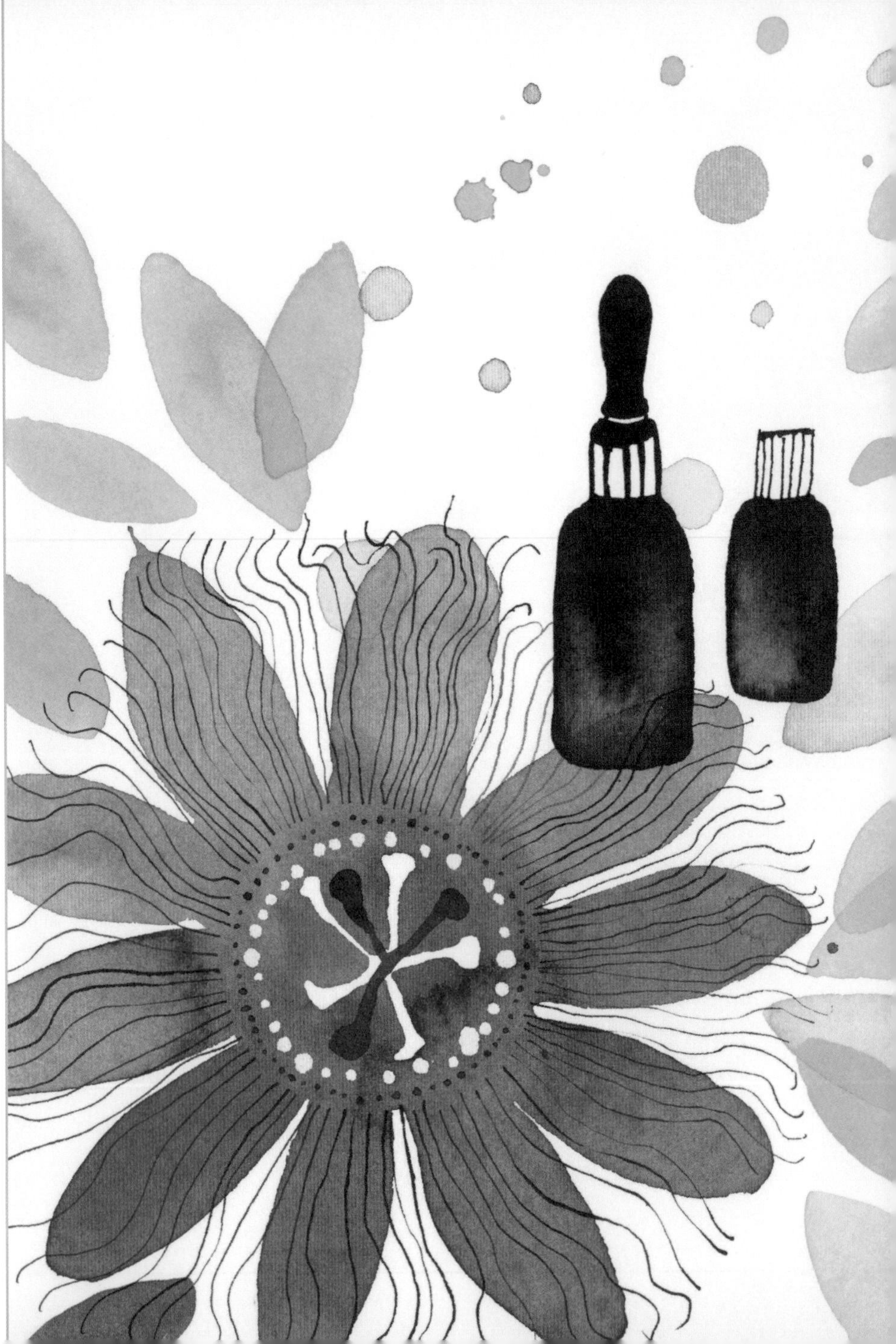

# HERBAL REMEDIES

Plants and herbs have been used as medicines for thousands of years, and scientific research confirms that many have active ingredients that are proven to aid relaxation and sleep.

Natural remedies are most often available as teas, tinctures (extract preserved in alcohol), and capsules or tablets from health shops, pharmacies, or supermarkets. As with all medicines, they can interact with other medications, so always tell your doctor about any preparations you're taking.

**Valerian** *(Valeriana officinalis)* is probably the best-known herbal sleep remedy. Taking 300–600mg in tablet form an hour before bed can increase your sleep length and quality. Valerian tea is a popular bedtime drink.

**St John's wort** *(Hypericum perforatum)* is a shrub whose therapeutic qualities are well-documented. It lifts the mood and can be used for insomnia and the anxiety and depression associated with it.

**Passion flower** *(Passiflora incarnata)* The sedative properties of this plant aid relaxation and promote sleep. It works by increasing your levels of gamma-aminobutyric acid (GABA), a compound that slows brain activity.

**Lavender** *(Lavandula angustifolia)* Essential oil can be used in a burner, room spray, or massage oil; dried flowers make tea or a scented pillow. Lavender soothes anxiety and improves sleep quality.

**Chamomile** *(Matricaria recutita)* is a gently calming herb, especially popular as a pleasant-tasting tea.

**Lemon balm** *(Melissa officinalis)* reduces anxiety and can ease breathing, which can be helpful at night if you have hayfever or asthma.

# YOGA SLEEP ROUTINE

Yoga can be an excellent sleep aid. It quiets the mind, relaxes the muscles, regulates breathing, and releases tension. There are several branches of the discipline – Hatha and Nidra yoga are both especially helpful for promoting sleep. Perform this simple routine before bed to encourage a restful state of mind, body, and spirit.

## 02

**Lying butterfly** Lie on your back, arms by your side. Draw up your legs and bring the soles of your feet together, so your legs fall to each side. Put pillows under your knees if you need to. Hold the pose for 30–60 seconds.

## 01

**Mountain pose** Stand with feet together, spine tall, and arms by your side. Lift your chest and breathe deeply, with your head and spine in line. Hold the pose for 30–60 seconds.

**NEED TO KNOW**

**BENEFITS** Improves quality and quantity of sleep; can help to reduce the need for sleep medication.

**TIME** Around 5 minutes. Practise every night before bed for maximum benefit.

## 03

**Legs up the wall** Lie on your back. Place your legs up against a wall so that your body forms an "L" shape. Hold the pose for 30–60 seconds, focusing on your breathing and feeling your whole body relax.

## 04

**Corpse** Lie on your back on the floor, with your legs relaxed and slightly apart and your arms by your side with palms facing upwards. Focus on breathing gently for 30–60 seconds.

# HOLISTIC MASSAGE FOR BETTER SLEEP

Holistic massage aims to enhance a person's emotional, spiritual, mental, and physical wellbeing. Many people find that massage is a key tool in achieving better sleep patterns.

All massage is relaxing and releases sleep-inducing hormones. Holistic massage, by treating mind, body, and spirit rather than any one specific symptom, may be especially helpful in combating specific sleep issues.

The positive physical effects you may expect, all of which can help improve rest, include:

• **Pain relief** as tense muscles are unknotted and aching joints become looser and move more freely.

• **Easier breathing** as your lung capacity increases and blood oxygenation improves.

• **Fewer toxins** – there is evidence that massage can help to drain the lymphatic system, your body's waste-disposal network.

Touch is all-important in holistic massage; it boosts levels of seratonin, a key hormone in promoting healthy sleep, and also improves self-esteem, promotes a sense of balance in body and mind, and relieves stress. Practitioners believe that emotions are liable to get "stuck" in our muscles and tissues, and can be released through physical manipulation.

Holistic massage is often combined with aromatherapy (see pages 88–89) – therapists use specific essential oils to boost the beneficial effects of the massage, and also to alleviate any particular problems.

If you decide to try holistic massage, ensure you consult a qualified therapist, who will be able to tailor a programme to match your particular needs. The more often you are treated, the more benefits you are likely to see. If you can, book your sessions after 6 pm, as close to bedtime as possible, so you are still enjoying the immediate effects when it's time for sleep.

# AROMATHERAPY

Aromatherapy is an ancient method of healing and relaxation based on essential oils extracted from plants and flowers. It can be remarkably successful in relieving sleep problems.

Aromatherapy uses essential oils to treat medical, emotional, and mental conditions. When absorbed by the skin or inhaled as aromatic vapour, the oils link with the memory and emotional centres of your brain to generate healing reactions. Just as importantly, they evoke happy memories and stimulate our pleasure centres. The result is the production of sleep-friendly hormone serotonin, as well as a range of other reactions, depending on the specific oil.

A range of oils can aid relaxation and sleep, including:

• **Lavender** A stress-reliever, it acts on the brain's neurotransmitters as an anti-depressant.

• **Sandalwood** Studies have shown that the beta-santalol compound in this rare wood has a sedative effect.

• **Bergamot** High in flavonoid compounds, bergamot oil acts as a muscle relaxant and anti-spasmodic.

• **Rose** This oil contains mild chloroform and ethanol elements that promote sleep and reduce stress.

• **Chamomile** This herb contains the sleep-inducing antioxidant apigenin that also acts on stress and anxiety.

## USING ESSENTIAL OILS

Most oils need to be diluted before use. Add a few drops to a warm bath for a soothing pre-bed soak. You can also fill a spray bottle with water and add oil to make a room or pillow spray. Added to a non-scented vegetable oil (known as a base oil), essential oils are excellent for massage and for dabbing onto pulse points or temples to ease stress and headaches.

Some essential oils are not suitable during pregnancy, or for people with certain medical conditions such as epilepsy or diabetes. Always consult a qualified aromatherapist for advice if you are unsure.

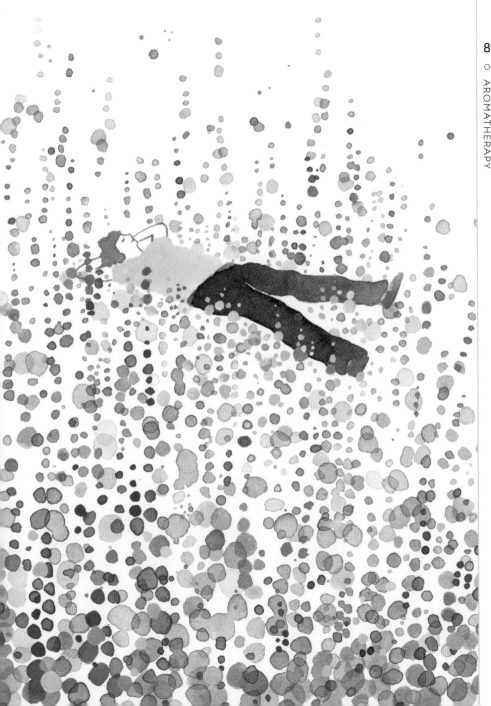

# SLEEP BETTER WITH ACUPUNCTURE

The aim of the ancient Chinese practice of acupuncture is to unblock, harness, and channel your life force, or *chi*, to boost wellbeing and remedy a range of physical, mental, and spiritual ailments, including sleep problems.

Acupuncture is the insertion of thin, sterile needles into the skin at key points of the body known as meridians. This action re-establishes the healing flow of *chi* energy around your system which in turn stimulates blood flow, relaxes muscles, and eases pain.

If you decide to visit an acupuncturist, ensure that they are licensed, with a proven background of training and particular interest and expertise in sleep problems. Sessions usually last around an hour, over a course of weeks or months; it can take at least 4–6 sessions before you notice any improvement in your condition. When assessing you, the practitioner will want to identify which one or more of these areas you will need to address together:

**Sleep onset issues**, which is essentially your inability to fall asleep once you go to bed.

**Sleep maintenance**, which covers the quality of your sleep.

" *Scientific research shows positive outcomes for those who undergo acupuncture.* "

**Intense, disruptive dreaming** – dealing with nightmares, which can lead to anxiety and sleep phobia.
**Insomnia** – the condition of chronic, health-threatening sleeplessness.

Practitioners treat sleep problems by focusing on particular meridians: the liver meridian for stress; the lung meridian for trouble waking up; the spleen meridian for dealing with worry-affected insomnia; and the heart meridian for deep-seated anxiety.

**MEASURABLE BENEFITS**
Scientific research shows positive outcomes for those who undergo acupuncture. An influential Canadian study in 2004 showed that acupuncture increased the body's production of melatonin, improved sleep times, and produced less disturbed sleep. Chinese clinical trials in 2018 and 2019 have found acupuncture to be more effective for insomnia than mainstream drug therapy, particularly for menopausal women.

# BREATHWORK FOR CALM

Breathwork – breathing mindfully and with control – is a powerful technique to balance the body, relieve stress, control anxiety, and boost self-esteem. It can also be very effective as a preparation for sleep. The art of bringing awareness to your breath is at the heart of meditation, yoga, and many other spiritual practices.

## 02

With your mouth slightly open and your jaw soft, gently take an in-breath. As you breathe out fully, contract the back of your throat slightly to make a gentle "Haa" sound.

## 01

Sit comfortably with your eyes closed and your back straight. If you like, you could play soothing music, or sounds of nature to set a relaxed mood.

## NEED TO KNOW

**BENEFITS** The action of focusing the mind on the process of breathing has a profoundly calming and grounding effect.

**TIME** 5–10 minutes, as you feel appropriate, just before bedtime.

## 03

Take a few more breaths like this. Then make the "Haa" sound on the inhale as well as on the exhale. Be aware of the gentle, unforced rhythm of your breathing.

## 04

As you breathe, visualize yourself sitting on a peaceful beach, breathing to the same rhythm as the waves flowing in and out on the shore. Continue for around 5 minutes then, when you are ready, return to your natural breathing.

# SLEEPING POSITIONS

If you have a favourite sleep position, it may not be helping you get the best rest. Knowing the pros and cons of each position will help you choose the right one for you.

**The foetal position** – on your side with knees bent – is one of the most popular sleep positions. It works well for most people, improving blood circulation and minimizing snoring. However, this position can make existing neck or back pain worse.

**The recovery position** is a variant of the foetal position; you lie on your side with your top leg bent over your straightened bottom leg. This position is good for reducing snoring, and may aid digestion. If you suffer from hip or back pain, placing a pillow between your knees will help align your spine.

**Sleeping on your back** is good for your back and joints, reducing pressure on them by distributing your weight evenly. However, it's the worst position for snorers and those who suffer from sleep apnoea (see pages 108–109).

**The log position** – on your side, with straight arms and legs – is good for snorers, and is beneficial for spinal health, as your body is well aligned. However, it can be uncomfortable for those with arthritis, and may cause pain in the hip of the top leg.

**Sleeping on your stomach** can help with snoring, but puts pressure on the spine, joints, and muscles, which can lead to long-term pain. It's especially hard on the neck muscles, which are turned to one side for hours at a time.

**Spooning** describes a couple facing the same way, with legs bent and close body contact. Spooning couples produce higher levels of oxytocin, the "love" hormone, but sleeping so close together can disturb sleep patterns. It's best to spoon for a while, then turn over to sleep in your own space.

# EXERCISE FOR BETTER REST

Our bodies are designed to move – and research shows that regular, moderate physical activity improves our mood and helps us sleep better and longer.

When it comes to exercise, every little definitely helps. Ideally, you should aim for around 30 minutes' moderate activity, five days a week, but if you can only manage 10 minutes, do it anyway – you will feel the benefit.

From a sleep point of view, it's usually best to exercise in the morning or afternoon. This is because exercise makes you hot, and in order to feel sleepy, your body temperature needs to fall. Evening exercise, while fine for general fitness, will impair sleep if you don't give the body enough time to cool down before bed.

## WHICH EXERCISE IS BEST?

Choose something you enjoy. You are likelier to stick with something that's fun – and consistency is key to improving sleep patterns. Bear in mind that if you are active outdoors, you gain the additional sleep-promoting benefits of daylight and fresh air, which boost serotonin levels to add to the energizing endorphins you are producing by exercising.

## REAP THE REWARDS

As well as improving the quality and length of sleep, exercise has other benefits. If you are overweight, burning more calories will aid weight loss, which in turn will help with some sleep conditions, such as sleep apnoea. And simply taking a break from the constant pressure of phone alerts and messages is excellent for managing stress and anxiety.

## EXERCISE CAUTION

Don't overdo it – over-exercising can actually cause insomnia. If you have any medical conditions, check with your doctor before embarking on an exercise regime.

# MANAGING YOUR MEDICATIONS

When taking prescription medication, always consult your GP or pharmacist about any side-effects you are likely to experience. It's very common for drugs to cause a sleep problem, or make an existing issue worse.

If you feel that a medication is impacting your sleep, don't tackle the problem yourself by changing or stopping your treatment, or by taking over-the-counter remedies in addition to the prescribed medicine – you need to talk to your doctor. There may be an alternative treatment you can try, or your dosage can be adjusted, or simply taking the medication at a different time of day can help considerably. Often, side-effects are temporary, and ease in a week or two.

**Anticonvulsants** are commonly used to manage conditions including epilepsy and bipolar disorder, but can cause daytime sleepiness and general lethargy. **Anti-arrhythmics** are prescribed to treat irregular heartbeat, but can cause daytime fatigue.
**Beta blockers** are routinely used for high blood pressure, but can cause insomnia, disrupted sleep, and even nightmares.
**Antihistamines** often make users feel sleepy. Non-drowsy options are available.

*" If you suspect your medication is causing your sleep problems, don't stop the treatment — see your doctor. "*

**Corticosteroids** are prescribed for a range of disorders, including asthma, allergies, lupus, arthritis, and Addison's disease. They suppress the immune system, but they also promote insomnia at night and restlessness by day.

**Diuretics** are prescribed to treat high blood pressure, but they also make the body produce more urine, resulting in broken sleep due to frequent nocturnal toilet visits, and leg cramps caused by night dehydration.

**Hormones**, such as the birth control pill, can affect sleep quality. A study found that women taking the pill tend to sleep less deeply than those who were not.

**SSRIs** (selective serotonin reuptake inhibitors) such as Prozac are used to treat depression, anxiety, bulimia, PTSD, and panic attacks. SSRIs break up sleep, and also make REM sleep harder to achieve.

## 01

An hour before bedtime, switch off all electronic devices and make sure they stay out of your bedroom. Dim the lights in your living room or switch on some lamps and quietly reflect on the events of the day.

## 02

When you feel sufficiently relaxed, start your pre-bed routine. Aim to do the same things, in the same order, every night: walk the dog, change into nightwear, brush your teeth – whatever you normally do before going to bed.

# WINDING-DOWN ROUTINE

We all need to decompress after a busy day, and for a good night's sleep it's essential, to dissipate the high adrenaline and stress-inflating cortisol levels we've built up during the day. Establishing a routine – and sticking to it – will help your mind and body to prepare for rest.

**03**

Now go back to your living room and do something you find relaxing; the aim is to bring adrenaline levels right down. You might read a few pages of a book, or try a repetitive, soothing activity or hobby such as knitting.

**04**

When you are ready, go to your bedroom, set your alarm and turn it away from you so you can't see the display. Get into bed, turn off the light straight away, and settle down with the intention of sleep.

**NEED TO KNOW**

**BENEFITS** Alleviates anxiety about not being able to fall asleep; a structured end to the day makes getting to sleep part of a predictable, soothing routine.

**TIME** Around an hour. Perform the exercise every night, until it becomes second nature. The more the ritual becomes ingrained, the less likely you are to fret about the possibility of sleeplessness.

# SPECIAL
# CIRCUMSTANCES

# MEDICAL CONDITIONS

A range of conditions affect sleep – from physical or mental illnesses to metabolic- or age-related problems. In addition to seeking medical treatment, there are lots of self-help measures we can take to minimize symptoms and get better rest.

# RESTLESS LEGS SYNDROME

If you often wake up at night with an itchy, crawling sensation in your lower limbs, you may have restless legs syndrome (RLS). Here's a rundown on what this nasty, sleep-disturbing condition is, why it happens, and how you can manage it.

This sleep disorder affects around seven per cent of adults. Women are more likely to suffer than men and there is a genetic component – it tends to run in families. Its onset is usually from middle age and it gets worse as you get older. RLS can also be an indicator of a separate condition, such as coeliac disease, so you should always consult your doctor if you experience symptoms.

## WHAT TO EXPECT

RLS is characterized by unpleasant itching, throbbing, and aching feelings in the legs. It happens after a period of inactivity, so occurs more before or during sleep. It can usually be temporarily relieved by moving or shaking the legs – hence the name. Episodes usually last from 15–40 seconds, and go on for an hour or two, but sometimes throughout the night. If you experience RLS once or twice a week it is classed as moderate; more than this is considered severe.

## MANAGING THE CONDITION

It's not known what causes RLS. There is no cure, but it can be much improved by lifestyle changes and medication. Iron supplements may help, as can avoiding alcohol, nicotine, and caffeine. Before bed, a warm bath or some gentle exercise can also help. Drinking a glass of milk provides sleep-promoting tryptophan (see pages 76–77). If medication is required, anti-seizure drugs are often prescribed. For severe RLS, stronger sedatives such as opioids may be necessary.

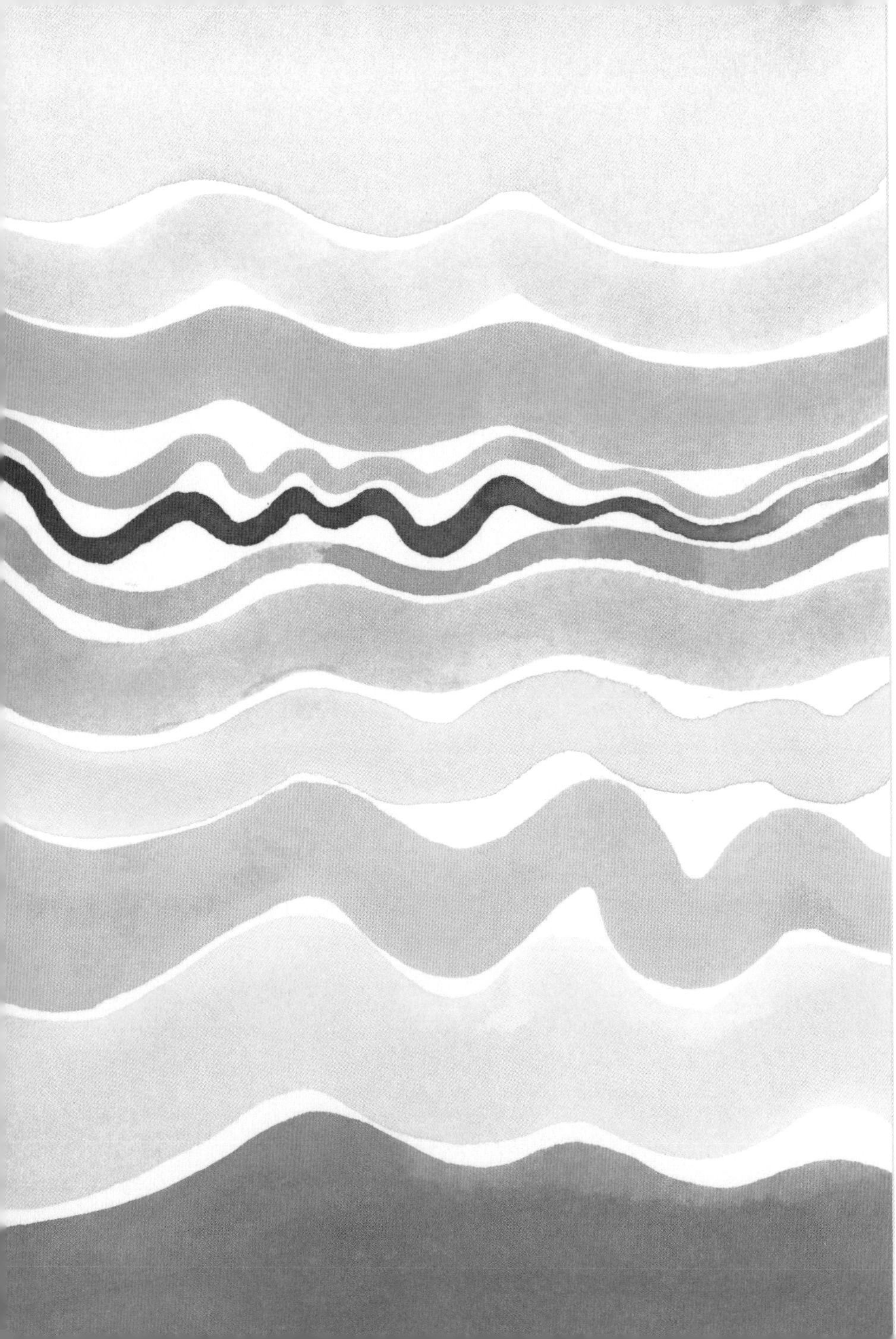

# OBSTRUCTIVE SLEEP APNOEA

Loud, sleep-disturbing snoring is the most obvious symptom of this sleep disorder, also known as OSA. It's a serious condition which, if left untreated, can cause significant health problems.

## WHAT IS OSA?

When we sleep, our throat muscles relax. If they relax too much, they can narrow or even cover the windpipe, stopping air from getting in, and causing us to wake suddenly, gasping for air. This is obstructive sleep apnoea, and in the most severe cases it can occur every 1–2 minutes throughout the night. Men are more susceptible to it than women, and it's more common after the age of 40. It can also run in families. You are more prone to sleep apnoea if you are obese, take sedative medication, drink a lot of alcohol, or have nasal problems such as a deviated septum.

Sufferers often don't know they have OSA, as the waking episodes last just a few seconds. They often feel worn out and sluggish the next day and wonder why. It's often partners who spot the condition first, being on the receiving end of the symptoms of laboured breathing, snoring, and gasping.

" *Lifestyle changes such as stopping smoking or losing weight can alleviate the symptoms of sleep apnoea.* "

## DEALING WITH THE ISSUE

As well as fracturing sleep, sleep apnoea can lead to other health issues, such as high blood pressure, strokes, and heart arrhythmia. So if you suspect you might have sleep apnoea, it's vital to see your doctor as soon as you can. You may be referred to a specialist sleep clinic to confirm the diagnosis.

**Lifestyle changes** can help alleviate symptoms: losing weight, cutting out alcohol or cigarettes, and sleeping on your side are all proven to reduce the severity and frequency of episodes.

**CPAP** (continuous positive airway pressure) is a small pump that delivers a continuous supply of compressed air through a mask that you wear while you sleep. The air keeps your airway open.

**MAD** (mandibular advancement device) is mostly used for milder OSA. It's a kind of gumshield that holds your jaw and tongue forward to increase the space at the back of your throat.

# BEATING INSOMNIA

It's common to call any temporary sleep issue insomnia, but it is actually a chronic condition of sleeplessness that can crush your quality of life. However, insomnia needn't be a life sentence; with the right help, most people overcome it successfully.

Lack of sleep or disrupted sleep crosses over into insomnia when it occurs regularly over a number of months, or even years. Insomnia takes a number of forms. If you experience one or more of the following, you should consult your doctor:

• You often or always lie awake and simply cannot get to sleep.

• You often or always wake in the night or very early morning and can't get back to sleep.

• You sleep very lightly, waking up constantly at the slightest disturbance.

• You feel debilitated and tired by day and need to take frequent naps.

• You suffer low mood or mood swings by day and find it hard to concentrate or focus.

## TACKLING THE TRIGGERS

Although insomnia requires specialist treatment, you can help yourself by minimizing any aggravating factors.

• To manage depression or counter stress and worry, practise mindfulness, relaxation techniques, or meditation. Expose yourself regularly to serotonin-producing sunlight.

• Consider any lifestyle changes you could make to boost your physical health. Could you exercise more? Eat more fibre and less sugar? Lose a little weight? Even small changes can make a real difference.

• Could you improve your sleep environment? For instance, a mattress that's more than 10 years old almost certainly needs replacing.

• Stay positive. It may take a while, but insomnia is a curable condition. Trust that you will sleep well again.

# UNDERSTANDING SLEEP PARALYSIS

Have you ever woken up unable to move? Did you feel
a sense of unease that you couldn't quite fathom?
You may have had an episode of sleep paralysis.
Don't worry: it isn't serious, and you are not alone.

## WHAT IS SLEEP PARALYSIS?

This unsettling phenomenon happens
if you wake up during an REM (rapid
eye movement) phase of sleep. REM
sleep is when you dream most, and
your nervous system as a consequence
"freezes" your muscles to stop you
acting out those dreams. So, if you
wake up suddenly during an REM
phase you feel paralysed until your
muscles "catch up" with your brain.

Episodes usually last from a few
seconds to a minute or two at most –
though the unpleasant sensation can
make it feel much longer. Research is
ongoing, but it is not believed to be
caused by any underlying emotional
or psychological issues.

## COPING STRATEGIES

As disconcerting as sleep paralysis can
be, it is not harmful. It can happen at
all ages, but most commonly affects
children and teenagers, possibly
triggered by rapid changes in their
developing brains.

You can't prevent sleep paralysis,
but there is evidence that good sleep
hygiene helps (see pages 24–25). It
doesn't appear to seriously affect sleep
patterns – although current thinking
is that it may be a sign that your body
is not moving evenly enough through
the sleep cycle, and needs to get back
in phase. See your doctor if the
condition makes you very anxious or
if episodes occur regularly.

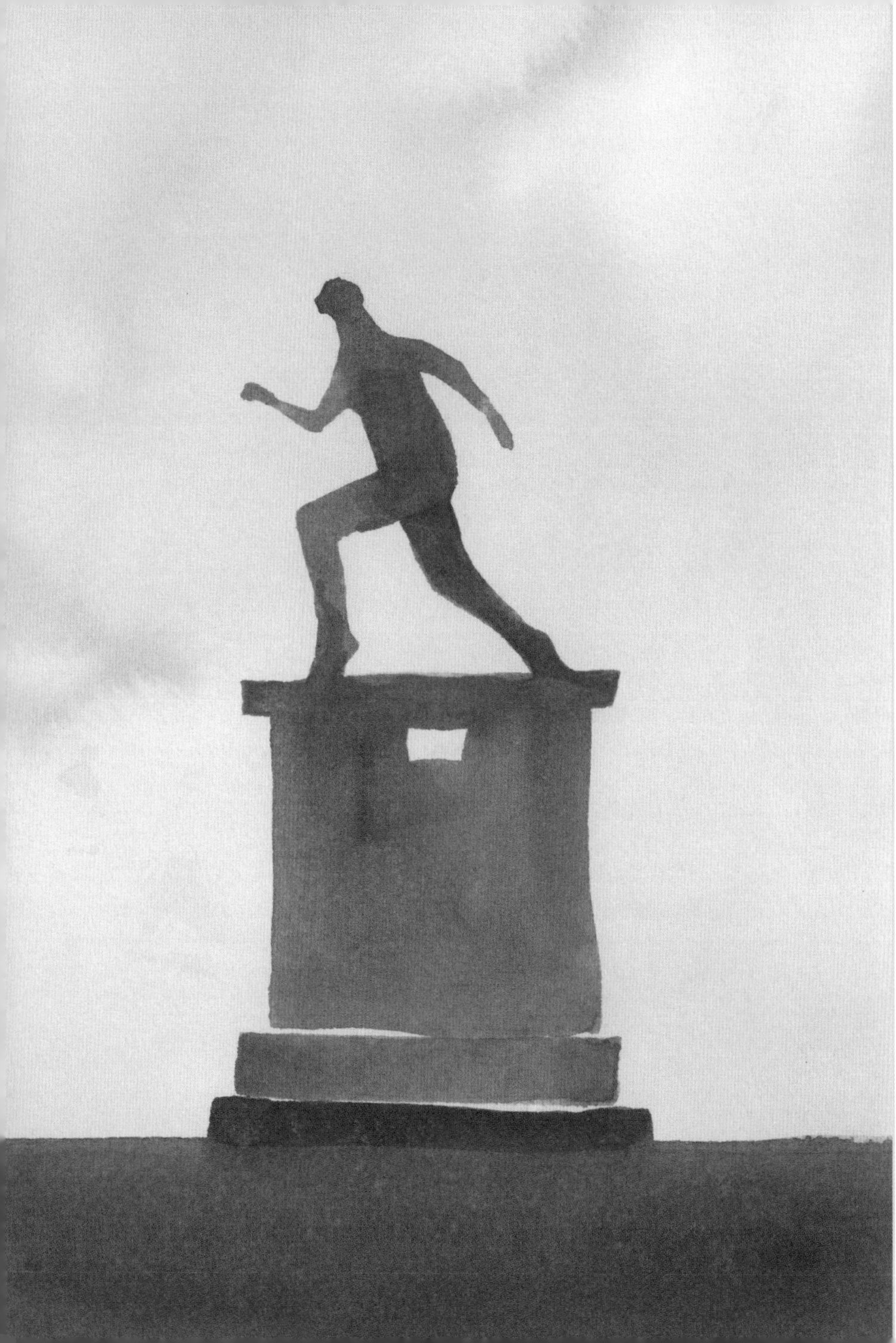

# LIVING WITH NOCTURIA

It's no fun constantly waking up at night for toilet visits. The condition is called nocturia and it's more common as you get older. Fortunately it can be treated, and you can also make lifestyle changes to alleviate symptoms.

Nocturia in adults is usually age-related. As you get older your bladder muscles grow slack and harder to control. Also, with age comes a range of conditions that impact on urination. If your habits change, always see your doctor. In most cases, effective treatments are available.

**Enlarged prostate**  The prostate gland, part of the male reproductive system, often enlarges with age. The prostate surrounds the urethra, the tube through which urine passes, and also presses against the bladder. This means that an enlarged prostate both restricts urine flow through the urethra and presses on the "fullness" sensors in the bladder walls. This leads to what feels like an almost constant need to go to the toilet, day and night, often with little or no urine to pass.

**Hormonal changes**  One function of ageing is that your body starts to produce fewer anti-diuretic hormones, which reduce your urine production, and in turn your need to urinate. Depleted levels of these hormones inevitably mean more frequent night toilet visits.

" *Maintaining a healthy weight means less pressure on the bladder, reducing the urge to go.* "

**Urinary tract infections** Mostly caused by bacteria entering the urinary tract or bladder, these infections make urination painful, hard to control, and irregular, as well as causing an inability to empty your bladder completely when you urinate.

**Heart conditions** A significant side-effect of heart disease is water retention. The water can accumulate in the tissues, particularly the lower legs, by day. When you lie down at night, this water is released into your system, finishing in the bladder and causing a need to urinate.

**Diabetes** Undiagnosed or uncontrolled diabetes makes you drink and urinate more. High blood sugar also irritates the bladder, increasing the urge to urinate.

**SELF-HELP MEASURES**
Try to consume your daily liquid intake by 8 pm. In the evening, avoid diuretic foods such as celery, watermelon, and cucumber. Diuretic tablets and support stockings can help with daytime water retention. Maintaining a healthy weight means less pressure on the bladder, reducing the urge to go.

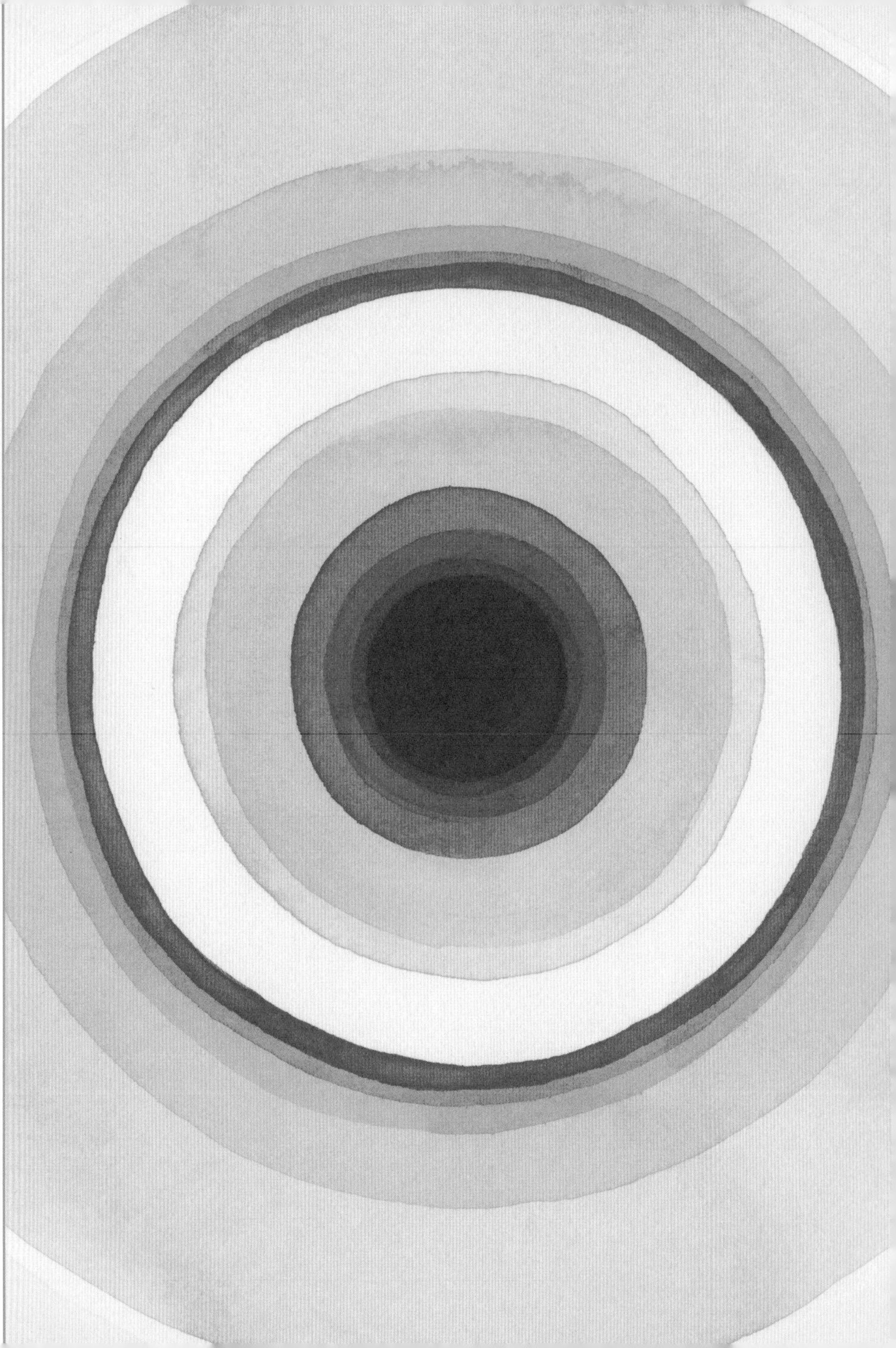

# COPING WITH CHRONIC PAIN

People living with chronic conditions such as arthritis, back problems, and migraines suffer more insomnia than the general population – and pain is usually to blame.

Chronic pain can disturb your sleep. In turn, the sleep deprivation aggravates your condition, generating more pain, creating a vicious circle. Your GP can advise on medication or recommend treatments such as physiotherapy, but you can also take steps yourself towards better sleep.

• **For muscle or bone pain**, take precautionary steps: arrange extra pillows to support painful joints, for example. A warm shower or bath before bed will relax muscles.

• **Relaxation techniques** can help distract your mind. Many people find that meditation (see pages 58–59) or breathwork (see pages 92–93) can help them to transcend pain. Visualization can help too, transporting you away from the stressful here-and-now to a more peaceful, pain-free place.

• **Complementary therapies** such as acupuncture, hypnotherapy, and holistic massage can help to break the pain/sleep cycle. Make sure practitioners are suitably qualified.

• **Good sleep hygiene** (see pages 24–25) can help, such as keeping to a regular routine, switching off mobile devices an hour before bed, and having a cool, dark sleeping space.

• **If pain wakes you** and you can't get back to sleep, get up and go into another room. Practise relaxation or read in soft lighting until you feel sleepy enough to go back to bed.

• **Recent research shows** that if you believe your pain will stop you from sleeping, the chances are that you will be right. Staying positive about your ability to sleep well gives you a much better chance of a restful night.

# MANAGING MENOPAUSE

Every woman's experience of the menopause is different, but sleep problems are among the most common and troubling side effects that women experience.

During the menopause, women's bodies produce less progesterone and oestrogen, hormones associated with calmness, relaxation, and the processing of serotonin. This leads to a variety of sleep-affecting symptoms:

• Hot flushes and sweating. These affect up to 85 per cent of menopausal women. An episode lasts for about three minutes, and at night, will seriously affect sleep.

• Snoring and sleep apnoea.

• Anxiety, depression, and mood swings, which can lead to insomnia.

• Joint aches and pains.

• Nocturia (see pages 114–115).

• Restless legs syndrome (see pages 106–107).

The main medical treatment for menopausal symptoms, including sleep disorders, is Hormone Replacement Therapy (HRT). Although HRT is effective, there are possible side effects and you will need to be closely monitored by your doctor throughout treatment.

**SELF-HELP SUGGESTIONS**
Whether or not you take HRT, there are practical steps you can take to help you sleep better:

• Use soya products, such as tofu and soya milk. These contain the plant hormone phytoestrogen, which can be helpful in reducing symptoms.

• Avoid overheating at night by keeping the bedroom cool and wearing loose nightwear, preferably in a natural fabric.

• Valerian tea before bed is excellent for promoting sleep, and users report that it is also effective in countering night hot flushes.

• Relaxation techniques, such as meditation, massage, and gentle exercise can reduce the stress that exacerbates symptoms.

# FEMALE HEALTH
# AND SLEEP

Some conditions specific to women can impact the ability to
sleep well. Here's how to avoid, remedy, or ameliorate
some of these effects.

**Premenstrual Syndrome (PMS)** This
affects up to 90 per cent of women at
some point. Symptoms include
headaches, cramps, and bloating. A
drop in oestrogen and progesterone
levels before your period can cause
low mood and wakefulness.

In the week before your period is
due, minimize symptoms by avoiding
alcohol, taking gentle to moderate
exercise, and eating oestrogen-rich
foods including flaxseeds, soy, nuts,
and fruits such as oranges or peaches.
Red clover supplements, available
from health shops, can also help.

Other health issues exclusively or
mainly affecting women include:

**Cystitis** This bladder inflammation is
more common in women, and can
keep you awake with pain and frequent
trips to the toilet. You may need
antibiotics; self-help measures include
drinking plenty of fluids and avoiding
scented soaps and shower gels.

**Thrush** A fungal infection that is much
more common in women, thrush
causes pain, itching, and a white
vaginal discharge. Fungicidal
treatments are available over the
counter; employ the same self-help
measures as for cystitis.

**Fibromyalgia** A chemical imbalance
causing widespread pain, headaches,
and irritable bowel syndrome (IBS), this
condition affects sleep significantly.
Women are seven times more likely to
experience it than men. Good sleep
hygiene (see pages 24–25) can help,
especially taking gentle exercise and
avoiding caffeine and alcohol.

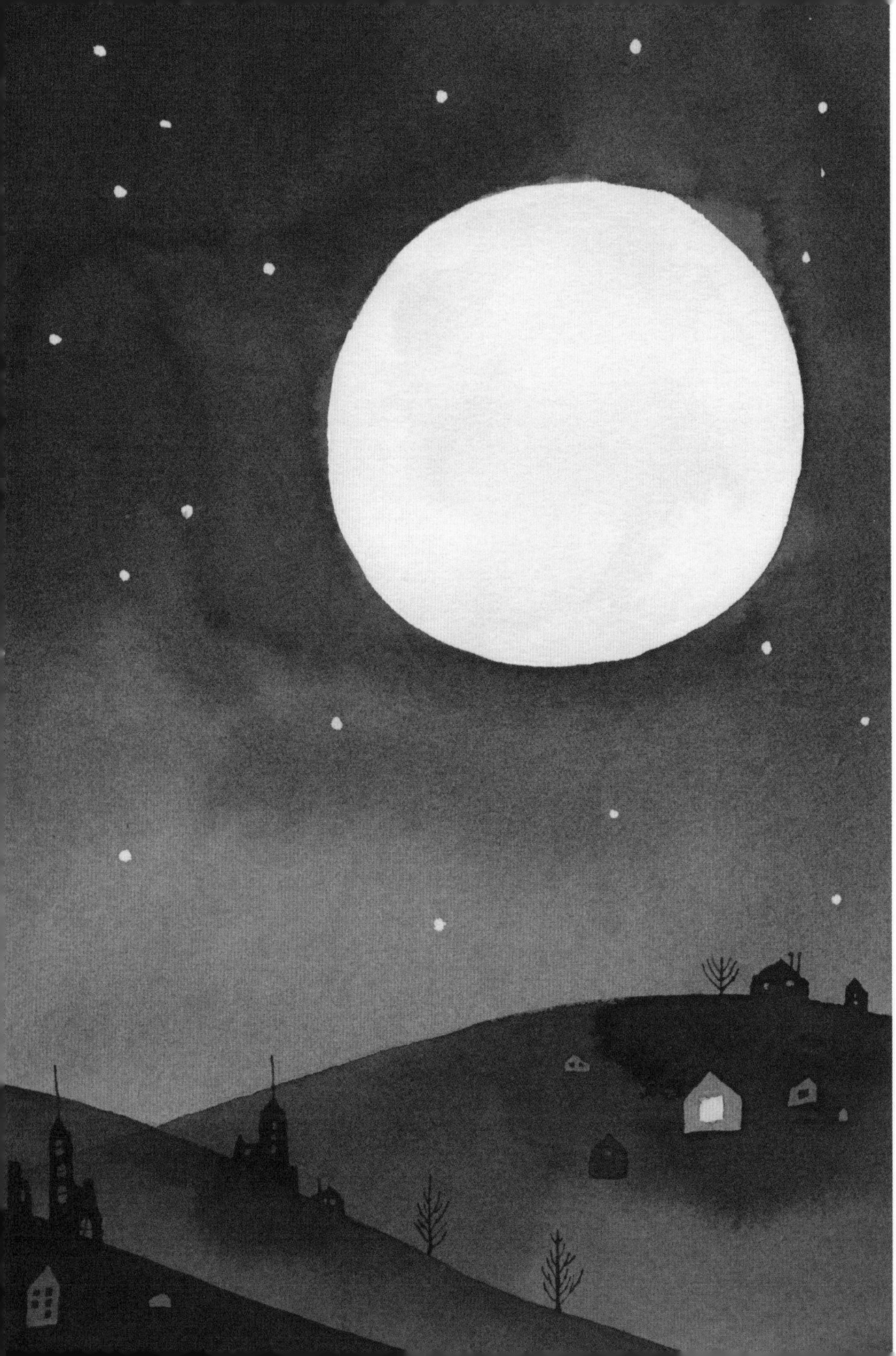

# HOW YOUR THYROID
# AFFECTS SLEEP

This small, butterfly-shaped gland in your neck produces
hormones to regulate body temperature and organ
function. A faulty thyroid can lead to a variety of health
problems, including disrupted sleep.

Women, especially those over 60, are
twice as likely as men to have thyroid
problems. These issues can be divided
into two categories:

**Overactive thyroid**, or hyperthyroidism,
is a condition in which the thyroid gland
produces too much hormone. Symptoms
include a rapid pulse, anxiety, mood
swings, and excess nervous energy. These
can all make sleep difficult, especially if
combined with night sweats, another
symptom of a thyroid working overtime.

**Underactive thyroid**, or hypothyroidism,
occurs when your thyroid doesn't make
enough hormone. Everything slows down;
you feel tired and sluggish, your brain is
fogged. You may become depressed.
This will compromise your ability to sleep,
as will the sleep apnoea (see pages
108–109) that often accompanies this
condition. Apart from the noise of
snoring, thyroid-influenced sleep apnoea
is characterized by a swollen tongue and
constricted throat – both of which can
affect your breathing while you sleep.

" *Women, especially those over 60, are twice as likely as men to experience thyroid problems.* "

## THYROID TREATMENTS

Both types of thyroid problem can be treated. For an underactive thyroid, a medication such as levothyroxine will be prescribed to top up your depleted hormone levels. This should rebalance your metabolism, resulting in more energy, reduced anxiety, and a much better night's sleep.

Overactive thyroid is treated with medication, radioactive iodine, or surgery, all of which reduce the thyroid's ability to produce the hormone. Each of these treatments has benefits and drawbacks that your doctor can discuss with you; however, the outcome for all three is usually an improved quality of life.

## LIVE HEALTHILY

There are lifestyle changes that can minimize sleep issues and other symptoms caused by thyroid problems. These include following a sleep routine (see pages 24–25), eating a healthy, vitamin-rich diet, and, for an underactive thyroid, taking gentle exercise.

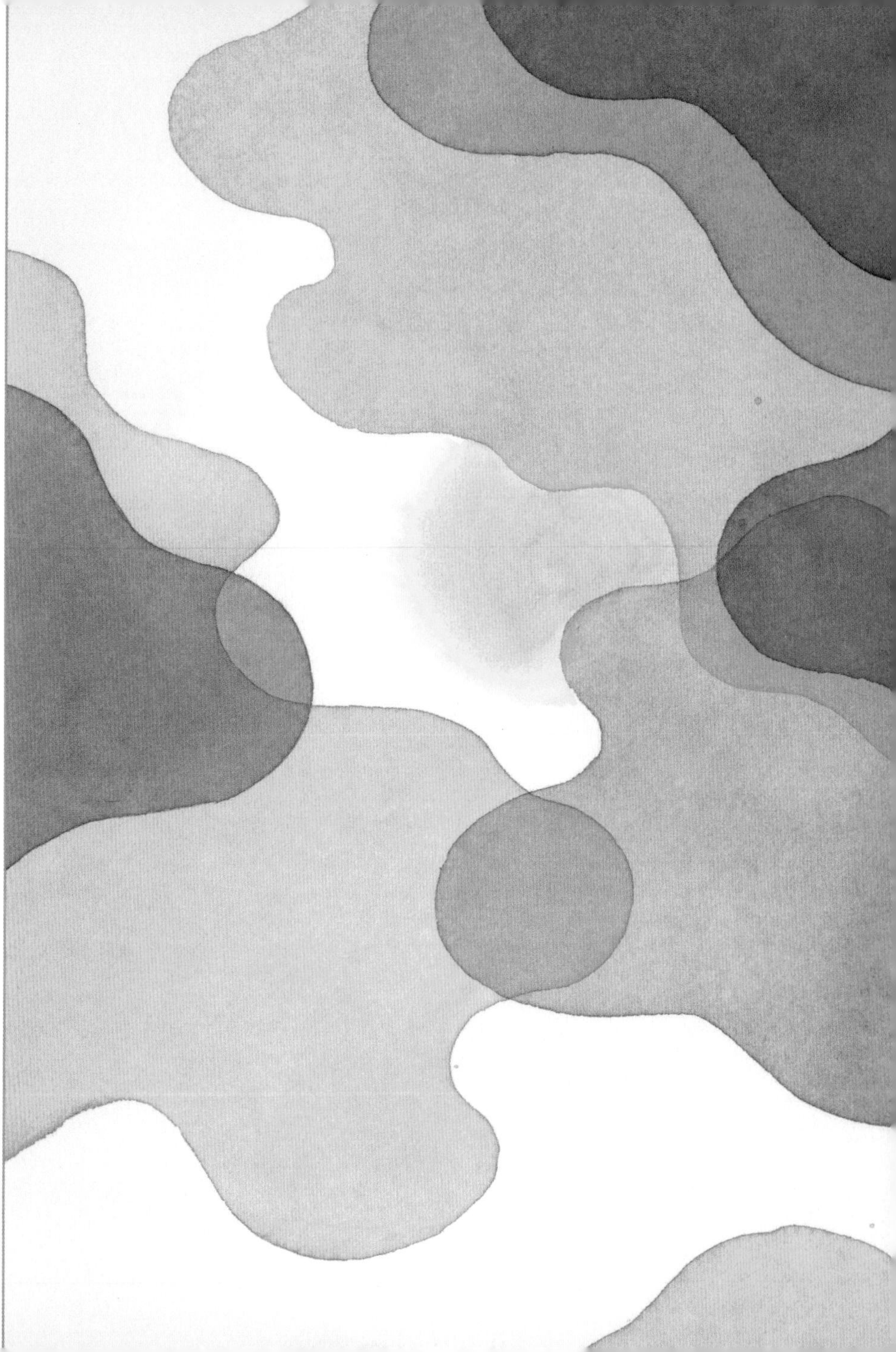

# MANAGING MENTAL HEALTH ISSUES

Poor sleep and mental illness feed off each other, setting up a vicious circle. Taking steps to improve your sleep can give you more focus and energy to manage and treat your condition.

The most common disorders that impact sleep include:

• **Depression** People with depression often fluctuate between not sleeping well and oversleeping, sometimes taking to their bed for days.

• **Seasonal affective disorder (SAD)** This form of depression is brought on by the lack of daylight in winter. Sufferers feel lethargic and low, and find it hard to wake up in the mornings.

• **Bipolar disorder** This causes swings between being very low and feeling "high" and overactive. Coping with such extremes can throw sleep patterns into chaos.

• **Anxiety** More than half of adults with anxiety disorder also experience panic attacks, obsessive compulsive disorder (OCD), and phobias. Sufferers are often unable to settle at night, with worries and preoccupations overwhelming even their bodies' physical need for rest.

## SELF-HELP OPTIONS

There are strategies you can try for better sleep, always in addition to your prescribed treatments. Never stop or alter your medication without talking to your doctor first.

• If your body clock feels out of sync, taking melatonin supplements may help you to rebalance.

• Light therapy, delivered via a special lamp called a light box, simulates sunlight and can encourage your body clock to reset itself.

• Exercise, especially aerobic activity, releases feel-good endorphins and sleep-friendly serotonin. Exposing yourself to sunlight at the same time will further boost the benefits.

• Relaxation and mindfulness often work well in partnership with prescribed medication. Try some of the exercises in this book, or enrol in a local yoga or meditation class.

# LESS COMMON SLEEP DISORDERS

A variety of sleep disorders, from the fortunately rare narcolepsy to debilitating chronic fatigue syndrome, can affect us during our lives. Treatments are constantly evolving, so it's important to seek your doctor's advice for any new or ongoing issue.

**Narcolepsy** Characterized by sudden, temporary collapse and loss of consciousness, this is a lifelong condition, although episodes generally decrease in frequency over time. It can be regulated with medication, but there is evidence that lifestyle changes can help, such as staying active, taking daytime naps, and ensuring good sleep hygiene (see pages 24–25).

**Chronic fatigue syndrome (CFS)**
Formerly known as ME (myalgic encephalomyelitis), symptoms include extreme tiredness, joint pain, dizziness, headaches, and palpitations. CFS is more common in women than in men. Treatments include pain- and sleep-regulating medication and CBT therapy (see pages 66–67). In most cases, the condition improves by itself, but this can take months or even years.

**Bruxism (teeth grinding)** This sleep behaviour can lead to headaches, jaw pain, earache, and damaged teeth. It's thought to be triggered by stress or anxiety; your doctor may recommend wearing a mouth guard in bed while the causes are addressed.

**Sleepwalking** This is one of a range of unusual night behaviours that clinicians term "parasomnia". It's more common in children, but may continue into later life. Treatment focuses on identifying and treating the underlying causes, such as stress or anxiety.

**Night terrors** These affect about 40 per cent of children, and involve screaming and thrashing about during sleep. They are usually triggered by similar factors to sleepwalking, with the same range of treatments offered.

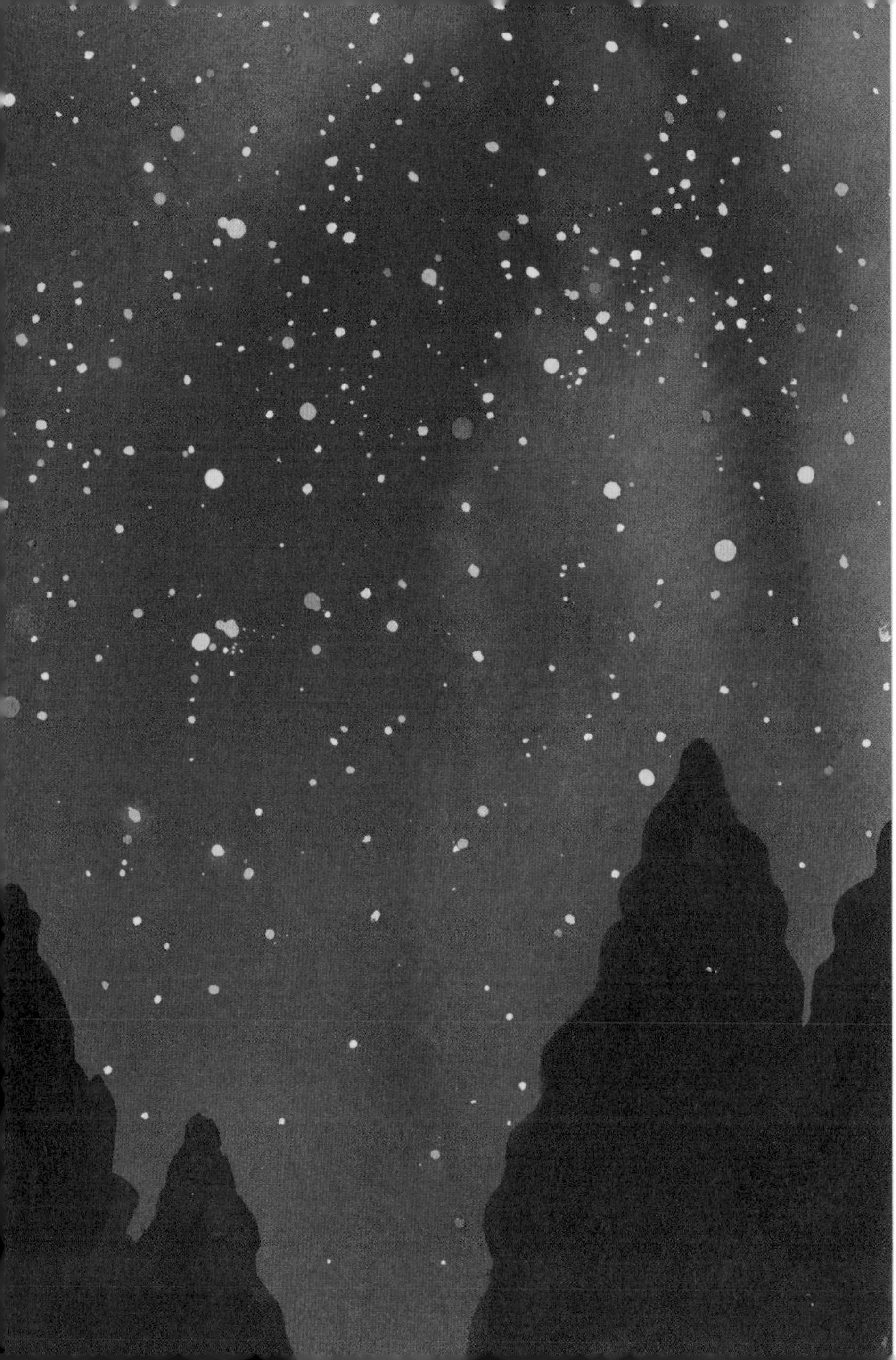

# DISRUPTED ROUTINES

Nothing ruptures sleep like having your routine turned upside down. But whether it's pregnancy, new working practices, or long and tiring travel arrangements, it's possible to resist these shocks to your system and take back control of your sleep.

# SLEEP BETTER DURING PREGNANCY

Sleeping well can be a challenge in the face of the physical and mental upheavals that pregnancy brings. However, there are plenty of ways to make sure you get the quality rest you need.

Up to 80 per cent of women report sleep problems during pregnancy. Symptoms that contribute to poor sleep include:

• **Snoring**, brought on by extra weight, and pregnancy hormones causing a blocked nose.

• **Nocturia**, the need to urinate through the night, caused by the uterus pressing on the bladder.

• **Nausea or heartburn**, caused mainly by hormonal changes.

• **Back or joint pain**, due to the extra weight you are carrying around.

• **Restless legs syndrome (RLS)**, possibly caused by lowered levels of iron and folate.

• **Disturbing dreams** that can break sleep, and are triggered by stress or hormone surges.

## SELF-HELP STRATEGIES

• Relaxation or meditation help to combat stress, which can make sleep problems worse.

• For heartburn, avoid spicy foods and triggers such as tomatoes. Over-the-counter antacids can help, but check with the pharmacist that they are safe.

• Eat smaller, more frequent meals so as not to overload your cramped digestive system.

• Eat folate-rich foods such as whole grains and cereals, or take folic acid and iron supplements.

• Try to sleep on your left side. This improves the flow of blood and nutrients to the foetus.

• Put pillows between your knees and behind your back to relieve pressure on joints and avoid back pain.

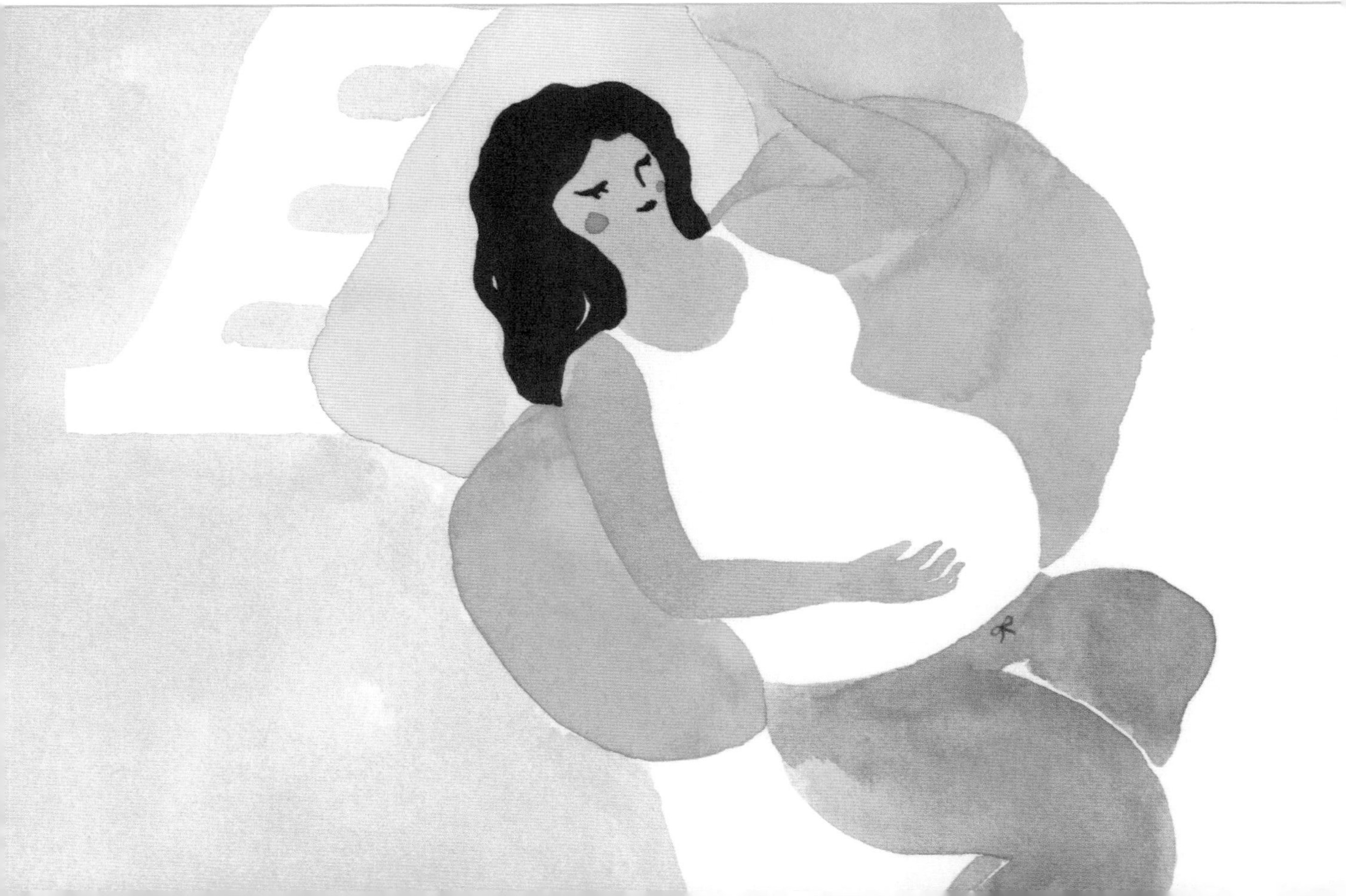

# UP ALL NIGHT, SLEEP ALL DAY

Not everyone goes to bed when it gets dark – there's an army of "night people", permanently or temporarily living out of sync with their natural body clocks. Here's how to survive and thrive on the night shift.

It goes against tens of thousands of years of genetic programming for us to live and work by night and sleep by day. But this is something many of us have to do, especially those who work in healthcare, transport, emergency services, communications, or service industries.

Currently, about 12 per cent of the UK workforce regularly works night shifts. Research shows that night workers tend to perform worse than daytime colleagues and are at increased risk of workplace accidents. What's more, they sleep less well and for less time during the day. Over time, people may develop shiftwork sleep disorder, which affects 10–30 per cent of shift workers. As well as causing tiredness, poor concentration, and depression, this condition also increases the risk of heart disease, obesity, and gastrointestinal problems.

## MINIMIZE THE EFFECTS

There are steps you can take to stay healthy if you have to be up all night.

**Prepare for shifts** Before you start night work, gradually nudge your sleeping and waking times later and later. This allows your natural circadian rhythms to synchronize to the new normal.

*It goes against tens of thousands of years of genetic programming for us to be up at night and sleep by day.*

**Keep regular hours** Stick to the same hours if you can. Evidence shows that rotating work patterns do the most damage to the body's internal clock.

**Eat less at work** Your digestive system does not adapt easily to eating in the middle of the night, so eat your main meal during daylight hours and stick to smaller snacks at night.

**Use bright light** When you wake up, expose yourself to bright light; this tells your body that its "day" is beginning.

**Take a nap** Before you start work, a quick nap (a maximum of 20 minutes) can boost energy and concentration.

**NEW PARENTS**

If you are caring for a baby, "night shifts" can follow an erratic pattern but try to stick to your normal times of going to sleep and getting up in the morning, no matter how many times you have to get up in between.

During night feeds, keep lighting low so as not to stimulate the waking-up hormones. In the daytime, take the opportunity to nap at the same time as your baby, but don't sleep for hours, even if the baby does – this can knock your body clock even more off-course and disrupt your night-time sleep.

# COPING WITH JET LAG

The upside of long-haul air travel is that we get to see more of the world than ever before; the downside is jet lag. It can play havoc with sleeping patterns and leave us tired for days, but there are things you can do to minimize the ill-effects of flying.

There are several strategies to try to reduce jet lag, before, during, and after a long flight:

• Adjust your sleeping habits a few days before you fly by one hour a day; earlier if you are flying east, later if flying west. This helps your body clock to adjust in advance of travelling.

• The day before you fly make sure you get lots of sunlight if possible and take a brisk walk. Sunlight promotes sleep.

• At the airport, set your watch to the time of your destination. This will give your body as long as possible to reset its internal clock to the new time.

• If flying west early in the morning, delay your sleep on the plane; if flying east to arrive in the morning, eat a high-carbohydrate meal at the airport (carbs can boost serotonin), and try to sleep straight away on the plane. Avoid alcohol while flying, as it can disrupt and affect the quality of sleep.

• On arrival, follow your new time zone's rhythms and routines as quickly as possible. Get as much serotonin-making sunlight as you can.

• Take melatonin supplements at night at your new destination, and for a few days after arriving home.

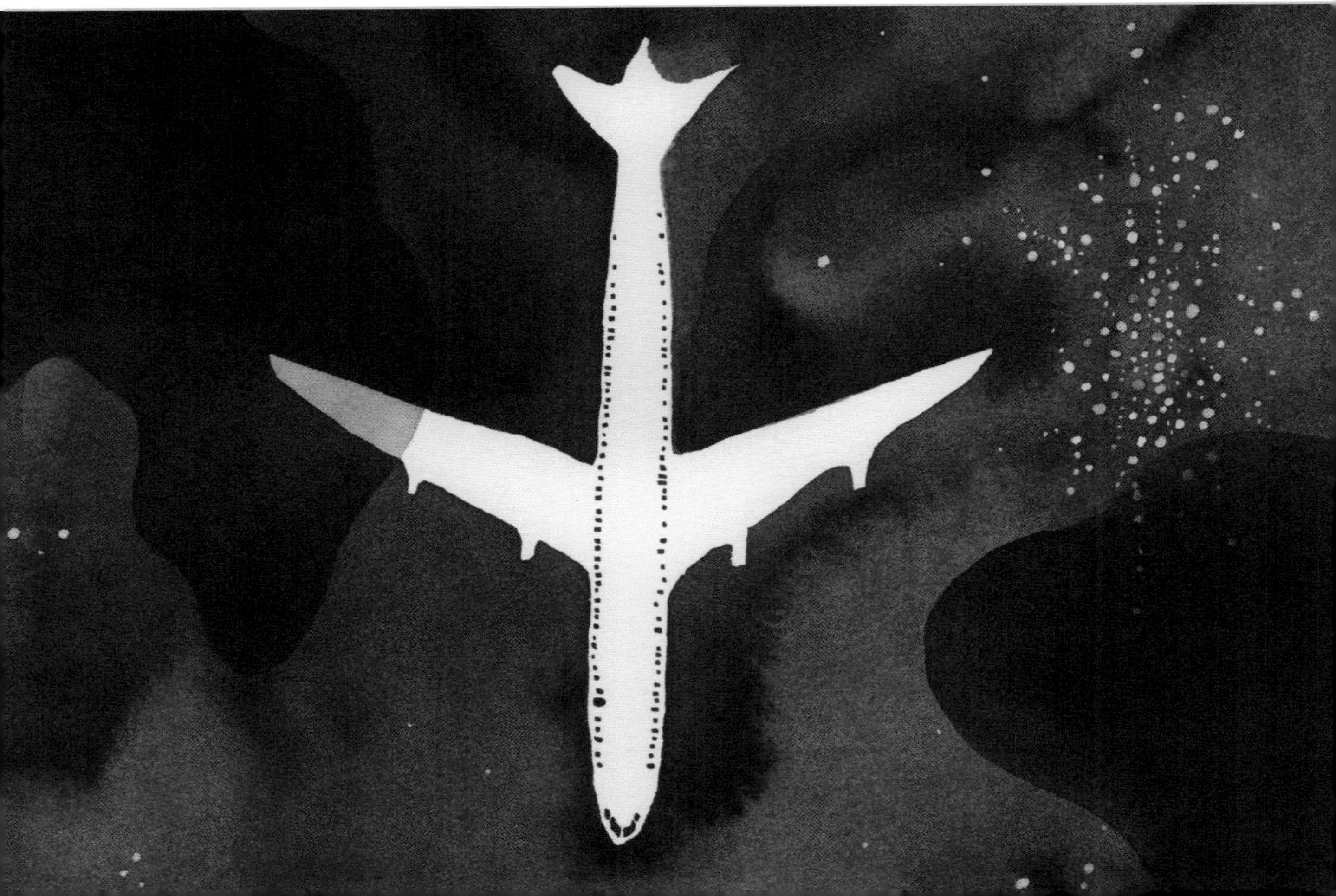

# DRIVING AND SLEEP

One-fifth of all motorway accidents are the result of tiredness. Here are the signs to look out for when you are fatigued, and some steps to take to avoid sleepiness at the wheel.

Driving tired is dangerous, both for you and other road users. If you have a sleep disorder, consider whether you're fit to drive – don't get behind the wheel if you have doubts about your ability to stay awake. In the UK, drivers are legally obliged to inform the licensing authorities if they have been diagnosed with conditions such as narcolepsy or sleep apnoea.

For a long drive, preparation is key. Here are some tips to get you safely from A to B:

• **Get proper sleep**, preferably eight hours, before your journey.

• **Avoid alcohol** at least 24 hours before you set off.

• **Stop before you feel tired**, and take a 20-minute break every couple of hours. Nap if you can, and it's safe to do so.

• **If you're driving** and experience any of the following, pull over as soon as you safely can: your eyelids grow heavy or you can't focus properly; you begin to daydream or your mind starts to wander; the car drifts and you clip the motorway rumble strip.

• **If you do get sleepy**, it's not enough to open the window for a blast of fresh air, turn up the radio, down a double espresso, or pop some caffeine pills. These fixes won't last more than a few minutes. You need to stop driving and rest properly before resuming.

• **Stay off the road** between midnight and 6.00 am if you can. This is when your natural circadian sleep rhythms are insistently telling your body that it's time to go to sleep.

# INDEX

# RESOURCES

Sleep is a relatively recent area of research worldwide, so it's important to get up-to-date information as advice may change as more is known. If you would like to try CBTi, check that the therapist is qualified in psychology or psychotherapy, as sleep problems almost always involve stress, anxiety, and depression and treating holistically is crucial.

## FURTHER READING

**Tired but Wired: How to Overcome Sleep Problems: The Essential Sleep Toolkit**
Dr Nerina Ramlakhan (Souvenir Press)

**Why We Sleep: The New Science of Sleep and Dreams**
Matthew Walker (Penguin Books)

**The Effortless Sleep Method** Sasha Stephens (Amazon)

**I Can Make You Sleep** Paul McKenna (Bantam Press)

**Herbal Stress Control: The complete and practical guide to using herbs to treat stress and anxiety** David Hoffmann BSc MNIMH (Thorsons)

**Reflexology and Acupressure** Janet Wright (Hamlyn)

## ONLINE

**www.londonsleepcentre.com**
**www.nhs.uk** UK information to find hospitals with sleep clinics
**www.evelinalondon.nhs.uk** Specialists in sleep disorders for children
**www.sleepassociation.org** Providers of sleep apps, some of which are free
**www.psychologytoday.com**
**www.binauralbeatsmeditation.com** Information on binaural beats/sounds
**www.theinsomniaclinic.co.uk** Sleep therapists throughout the UK
**www.sleepfoundation.org**
**www.sleepio.com** Provider of apps using CBTi techniques

## SOCIETIES/REGISTERS

| | |
|---|---|
| British Sleep Society | **www.sleepsociety.org.uk** |
| World Sleep Society: publishers of *Sleep Medicine Journal* | **www.worldsleepsociety.org** |
| American Academy of Sleep Medicine | **www.aasm.org** |
| The Sleep Council | **www.sleepcouncil.org.uk** |

## HYPNOTHERAPY

**www.hypnotherapy-directory.org.uk**

**www.general-hypnotherapy-register.com**

**www.nationalhypnotherapysociety.org**

## ACUPUNCTURE

**www.tuck.com/acupuncture-and-sleep/**

# ABOUT THE AUTHOR

Petra began her journey to becoming a sleep therapist with a diploma at the College of Stress Management in 1993, where she first trained in hypnotherapy. She opened her first practice while continuing to study, obtaining a PhD in Social Science.

Working as a hypnotherapist, she noticed that alongside resolving deep-seated psychological issues, many clients reported how much their sleep had improved after receiving therapy. It became clear that combining psychotherapy with hypnotherapy was a powerful combination, and Petra has since used this in her practice. Her work has taken her from the UK to the USA, Hong Kong, Germany, Denmark, and Ireland.

Following training in CBTi, Petra now practises as a sleep psychotherapist at the London Sleep Centre in Harley Street, where she continues to research and explore the many issues surrounding sleep.

# AUTHOR'S ACKNOWLEDGMENTS

It was both a surprise and an honour to be asked to write this book. As a new writer I received valuable guidance from the team at DK, so I thank them profusely for their patience; it has enabled me to learn a few tricks for the future! Many thanks to Dawn, Rona, Ian, Kiron, and their support teams, and to the illustrator and designers for the beautiful artwork in this book.

I would also like to thank every single client I've helped in the past, as they are the ones who have taught me the most. No matter what qualifications I have obtained, the real learning comes from observing the lives and problems of others.

The help and support from my beloved son James and my daughter-in-law Lynne have been invaluable. Thank you both so much for all your loving support.

I do hope the information in this book helps you to sleep better. Sweet dreams!

# PUBLISHER'S ACKNOWLEDGMENTS

DK would like to thank the following people for their assistance in the publication of this book: John Friend for proofreading, and Marie Lorimer for compiling the index.